SCIENTIFIC AND EDUCATIONAL JOURNAL OF THERAPEUTIC RIDING

2012/13

ISSN 1174-6238

ISBN 978-0-9845490-6-1

A PUBLICATION OF

THE FEDERATION OF HORSES IN EDUCATION AND THERAPY INTERNATIONAL A.I.S.B.L.

THE FEDERATION OF HORSES IN EDUCATION AND THERAPY INTERNATIONAL A.I.S.B.L

SECRETARIAT
HETI
P.O.Box 134
Damariscotta, ME 04543
Tel: +44 207 1831 256
Email: office@HETIfederation.org
www.HETIfederation.org

ISSN 1174-6238
ISBN 978-0-9845490-6-1

Table of Contents

Dear HETI Members

We are so pleased to have the 2012-13 Scientific and Educational Journal of Therapeutic Riding published and distributed to the membership. The HETI Executive Committee thanks you for your patience as we continue to pursue the goal of transforming the Journal into a peer-reviewed Journal. We have received a number of questions related to the Journal, its contents, and future publication dates. Hopefully we can answer all of your questions here.

Why did HETI choose to change the content of the Journal?

The field of equine assisted activities and therapies has grown tremendously in the last decade and it has become increasingly evident that dedicated research was necessary to continue moving forward.

It is HETI's desire to support that development by publishing a journal for those research findings and to give authors a forum that also satisfies academic requirements, making them eligible to receive grants. Peer review is an essential pre-requisite for this goal.

Why did it take so long for HETI to publish this journal?

Once the decision was made to broaden the scope of the Journal from a compendium of research articles, case studies, profiles, and general interest summaries, to a peer-reviewed scientific journal, it was necessary to identify an editor. We actually chose two editors – Nina Ekholm Fry and Susan Cotton who possess the credentials and knowledge necessary to begin this transformation (see the Editorial for their individual credentials). The editors were then tasked with defining the manuscript requirements and submission protocol. Only after that framework was in place, could they begin reviewing manuscripts that had already been received, put out calls for papers, network with researchers, and work with authors on the editing of their submissions. This was a time-consuming process.

Will non-scientific articles, informational pieces, and opinion summaries have a place in HETI publications?

Absolutely yes. We envision a split Journal in the future – a ranked, open access journal combined with an informational supplement. Please continue to send manuscripts for inclusion in either category, but please follow the submission process described on the HETI website.

When will the next Journal be published?

We anticipate publishing a Journal in 2014; however, we cannot publish what we don't have! Please encourage anyone you know to submit manuscripts and to do so as soon as possible. Please feel free to refer them to the editors if they have questions about the scientific or technical nature of the Journal.

It is our hope that you will continue to enjoy the Journal as a premier source for verified information about equine assisted activities and therapy.

Best regards in the New Year,

Ann O'Shallie
President, HETI (The Federation of Horses in Education and Therapy International A.I.S.B.L.)

Gisela Rhodes
Executive Director, HETI (The Federation of Horses in Education and Therapy International

Editorial

It has been a year of development for the *Scientific and Educational Journal of Therapeutic Riding* (SEJThR). We, Nina Ekholm Fry and Sue Maree Cotton, your new journal editors, have enjoyed the editing process and the articles in this issue. Most of the year has been spent on planning significant changes that will come into effect next year: a new journal structure, a peer-review system, and new submission guidelines.

We would first like to introduce ourselves. Nina Ekholm Fry is Associate Professor of Counseling Psychology and Director of Equine-Assisted Mental Health at Prescott College in Arizona, USA. She is an executive board member of the Certification Board for Equine Interaction Professionals (CBEIP) and trained in a number of equine-assisted methods. Her main interests in the equine-assisted (therapeutic riding) field are professional standards and training; equine welfare; research; and ethics. In addition to client work and teaching, Nina conducts equine-assisted facilitation workshops on a regular basis. She is a certified riding instructor and member of the Equine Research Network (EqRN). Sue Cotton is Associate Professor and Principal Research Fellow based at Orygen Youth Health Research Centre, at University of Melbourne, Australia. Sue has a background in clinical neuropsychology and biostatistics. At her centre, she has led equine-assisted therapy programs for young people who have had mental illness (e.g., psychosis and or depression). She is currently involved in numerous clinical trials of both pharmacological and psychotherapeutic interventions for young people with schizophrenia, bipolar disorder, borderline personality disorder, substance use disorder, and depression.

Our goal for the year ahead has been to create a journal structure that more closely meets the needs of our field: a forum where not only research, which is much needed, is presented, but where case studies and other practice-related articles can be shared. The new structure includes the following sections: Original Research, Review, Case Study, Theory and Practice, and Letters to the Editor. More information about each section can be found on the HETI website on the journal page: www.hetifederation.org/journal.html

The call for submissions for next year's journal is now open and we encourage you to take a look at the new manuscript submission guidelines on the HETI website. It is our intent to raise the standard of professional submissions to the journal and to ensure that the submission guidelines and process are as clear as possible.

In addition to the call for submissions, we are announcing a call for peer-reviewers as a step in the process of bringing the journal to a peer-reviewed status. Peer-reviewers conduct blind scholarly reviews of manuscript submissions within their field of knowledge. If you or someone you know is interested in serving the journal in this crucial way, please have a look at the call for reviewers in this issue (also on the HETI website).

We hope you will enjoy the articles in this issue of the journal. They include a discussion on science versus pseudoscience (*Hippotherapy: The Mysterious Art of Healing or the Art of Rehabilitation*), a case study on movement in patients with ADHD *(Quality of Movement and Motor Problems in Patients with ADHD),* and two studies; one on postural differences between riders and non-riders (*An Investigation into Seated Postural Differences between Horseback Riders and Sedentary Non-riding Individuals*) and one on psychological effects of therapeutic riding in two settings (*The Psychological Effects of Therapeutic Riding (TR) for Children).* We hope that you will share in our excitement for the changes taking place this coming year. As always, please send us an email with your comments and questions to editorial@hetifederation.org.

Sincerely,

Nina and Sue

Nina Ekholm Fry
Department of Counseling Psychology
Prescott College, Arizona, USA

Sue Cotton
Orygen Youth Health Research Centre
University of Melbourne, Australia

Call for Manuscripts

The Federation of Horses in Education and Therapy International (HETI) publishes the Scientific and Educational Journal of Therapeutic Riding (SEJThR) annually. We are currently seeking submissions with the purpose of advancing knowledge in the fields of therapeutic riding and equine-assisted activities and therapies.

The SEJThR has several submission categories:

Original Research
An original research study conducted by the author

Review
A comprehensive or systematic review of research on a specific topic

Case Study
A presentation of a single comprehensive case study or multiple case studies

Theory and Practice
A discussion of a specific topic in the field

Letter to the Editor
Comments on articles published in the SEJThR

Manuscripts are reviewed through a double-blind peer-review system and submissions are welcome throughout the year.

For more information about manuscript requirements and the submission process, please visit www.hetifederation.org/journal.html.

Please contact journal editors Nina Ekholm Fry and Sue Maree Cotton with questions.

Journal Editors:
Susan Maree Cotton • Nina Ekholm Fry
editorial@hetifederation.org
www.hetifederation.org

Call for Reviewers

The Federation of Horses in Education and Therapy International (HETI) publishes the Scientific and Educational Journal of Therapeutic Riding (SEJThR) annually. The focus of this journal is on advancing knowledge in the fields of therapeutic riding and equine-assisted activities and therapies.

Reviewers are needed to ensure the quality and status of the SEJThR as a peer-reviewed journal. Peer-reviewers conduct blind reviews of manuscript submissions within their area of knowledge as a step in the review process.

Expectations for reviewers are:

- Ability to provide constructive feedback for a scholarly manuscript in terms of its quality and contribution to the field
- Timely and respectful communication with journal editors

Requirements for reviewers are:

- Professional experience in an area of therapeutic riding/equine-assisted activities and therapies
- A master's or doctoral degree
- Experience in reviewing scholarly texts
- Ability to complete a review within four weeks of receiving a manuscript

If you are interested in contributing your time as a peer-reviewer for the SEJThR, please submit a letter of interest and your curriculum vitae (CV) per email to editorial@hetifederation.org. Please type "Reviewer Application" in the subject line.

ARTICLE ONE

The Psychological Effects of Therapeutic Riding (TR) for Children

Andrea Carey, MA candidate, York University
Sarah Murray, BA, BEd, Brescia University College
Anne Barnfield, DPhil, Brescia University College

Correspondence to:
Dr. Anne Barnfield, Associate Professor, Psychology,
Brescia University College, 1285 Western Road,
London, Ontario, Canada, N6G 1H2.
Phone: (519) 432-8353, ext. 28246
Email: abarnfie@uwo.ca

Support:
Dr. Benjamin Goldberg Research Award to A. Carey
Brescia University College Faculty Development Allowance (FDA) to A. Barnfield.

Acknowledgement:
The authors would like to thank the staff, volunteers, riders and their parents at SARI TR. We are most grateful for their aid with this research.

ABSTRACT

In this paper we present two studies which investigated the psychological benefits of therapeutic riding (TR) for children, primarily those with disabilities. The studies were run during two TR programs: Study 1, TR delivered in a summer camp; and Study 2, once-weekly TR sessions across two 10-week courses. All participants were tested at SARI Therapeutic Riding (SARI), London, Ontario, Canada. Survey methodology was employed, using standardized, validated questionnaires developed by the American Camp Association to make within-subjects comparisons. Data was collected at pre-camp/TR course, post-camp/TR course and follow-up time points; obtained from parent and child questionnaires, from researcher observation checklists, and through semi-structured interviews with the parents of children who attended summer camp. On all measures the participants of the camp demonstrated statistically significant gains between pre- and post-camp with respect to the domains of positive identity, social skills, physical skills, and positive values (all at $p<.01$ level). For weekly riding sessions, results from parent surveys were not statistically significant. Observation checklists completed by the researcher did show positive change in behaviours; one-way analysis of variance (ANOVA) comparisons of the domains measured in pre-, mid-, and post-course observations were significant, $p<.01$.Therefore it appears that TR, delivered in a summer camp or in a 10-week course, can be beneficial to children of all ability levels.

KEYWORDS

Children; Therapeutic Riding; Psychological Benefits.

INTRODUCTION

Thousands of children across North America attend summer camps and participate in sporting activities during the remainder of the year. Both activities are seen as of benefit to children of all abilities. A particular sub-group taking part in camps and athletic activities is that of children with special needs. Being able to join in activities with their more able peers is a goal for many such children and their parents. Camps serve as a place where children can learn how to live, work, and act in a community; learn and refine social skills; increase their self-efficacy; and transfer these skills to their everyday lives (Colyn, DeGraaf, & Certan, 2008). Participation in summer camps and similar programs has been shown to have a range of benefits for children, including improvement of sports skills and formation of positive social relationships (Siperstein, Glick, Bardon, Harada, & Parker, 2007) and fostering social and emotional competencies (Catalano, Berglund, Ryan, Lonczak, & Hawkins, 2004; Dworken, 2001). Henderson, Schueler Whitaker, Bialeschki, Scanlin, and Thurber (2007) launched a longitudinal national research project that investigated the effects of a week of summer camp on children using the Camper Growth Index questionnaires (Henderson, Thurber, Schueler Whitaker, Bialeschki, & Scanlin, 2006). The researchers reported significant growth in the areas of adventure/exploration, independence, friendship skills, positive identity, and peer relations. Additionally, it was found that parents observed improvement in secondary areas, including: leadership, positive values, decision-making, spirituality, environmental awareness, and social comfort. From such studies, it is evident that camp is a beneficial activity for children and youth.

Camps may be especially beneficial to children with disabilities since it has been shown that these children tend to experience more difficulties in establishing relationships with peers (Burt, Obradovic, Long, & Masten, 2000), have lower self-concepts and self-efficacy (Klassen & Lynch, 2007), show less participation in physical activity (Gendron, Royer, Bertrand, & Potvin, 2004), and have more difficulties in academic settings (Klassen & Lynch, 2007). Camps designed for children with disabilities enable campers to learn and improve new skills, overcome fears, form relationships, become self-reliant and independent, gain confidence, develop coping skills, increase their social comfort, improve self-concept and efficacy and feel 'normal' (Shaver & Scheibe, 1967; Michalski, Mishna, Worthington, & Cummings 2003; Rinn, 2006; Gaskell, 2007; Van Belois & Mitchell, 2009). Although it is evident that camps are beneficial for special populations, children with disabilities often have other challenges to overcome and have to incorporate therapy into their routines. Such therapy may involve physical therapists, occupational therapists and speech-language pathologists and typically takes place in an office setting; however, animal-assisted therapy (AAT) has recently become more popular, and has been found to be effective (e.g., Sterba, 2007). AAT, including animals to implement necessary treatment, can produce beneficial psychological, cognitive, social, and physiological outcomes (Beck, 2011; Esposito, McCardle, Maholmes, McClune, & Griffin, 2011).

Another branch of research has investigated the therapeutic effects of animals on people. A review by Wells (2009) discussed evidence that an animal could improve a person's health and well-being. It was reported that petting an animal could have positive short-term effects such as lowering blood pressure and lessening stress, and in the longer term that animals could lessen anxiety, loneliness, and depression and help increase feelings of autonomy, competence, and self-esteem in adults. Interaction with animals thus can provide many short-term and long-term benefits (Wells, 2009).

Although Ancient Greeks believed horses had healing powers, the first TR programs were not established until the mid-1900's (Bizub, Joy, & Davidson, 2003; DePauw, 1986), and differences of opinion still exist as to how therapeutic riding is defined. In the United States, possibly for reasons relating to medical insurance, there is a more rigid definition than in other countries. The Canadian Therapeutic Riding Association (CanTRA) website states that: "Therapeutic Riding is a form of therapy using horses to help persons with disabilities ..." (CanTRA, n.d.). Bass, Duchowny, and Llabre (2009) stated that TR is a subtype of animal assisted activity, in other words, not a therapy, "defined as using horseback riding treatment to improve posture, balance, and mobility while developing a therapeutic bond between the patient and horse" (p. 1261). In general, from studies which use the term TR, it seems most current research on this activity's effects has been in the area of physical disabilities, particularly cerebral palsy. TR can strengthen muscles used for walking due to the horse's rhythmic movement, and the warmth released from the horse's body has been used to relax a rider's tight leg muscles (Davis et al., 2009; Lehrman & Ross, 2001; Snider, Korner-Bitensky, Kammann, Warner, & Saleh, 2007).

From such findings it appears that TR has positive effects on individuals with physical disabilities, but less research has been dedicated to its effects on individuals with intellectual disabilities. The few studies conducted have reported some benefits, for example TR can teach individuals how to care for another being (the horse) and this, in turn, can help with the development of other healthy relationships (Karol, 2007; Bizub et al., 2003; Corring, Johnston, & Rudnick, n.d.). Caring for the horse may also lead to increased individual responsibility, autonomy, and awareness (Karol, 2007; Corring et al., n.d.). TR is sometimes said to be more effective than traditional therapy because the horse provides touch, contact, and rhythm that could not be provided in an office and the client can become a life-long participant in his or her habilitation through enjoyable recreation (Sterba, 2007). Thus, TR can be used with children with special needs to reach goals across a range of domains: cognitive, social, physical, educational, and behavioural (Dranch, O'Brien, & Kreger, 2010; American Youth Horse Council and Pennsylvania State University, 2006).

It appears that TR can have positive effects for children with special needs in social and motor functioning domains. To date, however, there has been little research investigating psychological effects of TR for children with disabilities. The current research was conducted to examine the effect of TR on children with physical or intellectual disabilities in four main areas of development: positive identity, social skills, physical skills, and positive values. Two types of TR programs were offered: (i) a camp environment; and (ii) once-weekly TR sessions across two, 10-week courses.

It was hypothesized that both a summer camp experience and TR course sessions would show benefits for children with disabilities. Specifically, it was expected that improvements in positive identity, social skills, physical skills, and positive values would be observed.

METHOD

PARTICIPANTS

All participants were recruited from SARI Therapeutic Riding, a non-profit centre for TR provision located in London, Ontario, Canada. SARI provides TR for children and young adults with special needs/disabilities (Note: both terms are used in this article, as employed by other authors and organizations. Disabilities can be seen as a more overall term, special needs as referring to more specific issues and outcomes). Participants were recruited using a convenience sample; i.e., those who signed up to participate in the sessions were invited to take part in the study. Beyond SARI's own regulations there was little that could be done to specify exact criteria for the participants in these studies, and to set very specific limitations would render the research impractical. Furthermore, the research was meant to address the impact of TR across all of the individuals who are eligible to enroll in TR programs such as the one offered by SARI, and so we intentionally attempted to include a broad range to more closely reflect the actual participants in TR programs.

Study 1: Camp program. Any child who was signed up for at least one week of summer camp in July or August was eligible to participate in the study. In the summer, weekly camps take place from 8 am to 4 pm, Monday through Friday, for eight weeks with a maximum of 12 campers per week. In the morning, the children would attend their TR sessions. The usual format for TR sessions was two children, each mounted on an individual pony/horse, in a semi-private session. Instructors were certified to CanTRA CTRII level. Level II instructors are "certified to teach up to four students at a time, giving safe and enjoyable riding instruction to children and adults with disabilities" (CanTRA, n.d.). In the afternoon, the campers participated in special activities that varied each day, e.g., arts and crafts, swimming, etc. Children attending the camp were either of typical ability or with a disability (e.g., Angelman's syndrome, Down syndrome, global developmental delay, cerebral palsy, autism spectrum disorder), as reported to SARI by parents.

Study 2: Weekly sessions. Children who had not previously participated in TR at SARI, and who had signed up for a 10-week riding program with weekly sessions, were eligible to participate in this study. Riding sessions took place Monday to Saturday from 10 am to 8 pm and children were assigned one 30-minute TR session per week within these days/times. The TR sessions were conducted identically to those during the summer camp. All children who participated in this study had special needs (e.g., cerebral palsy, autism spectrum disorder, Down syndrome, learning disability), as reported by parents.

For both studies, all TR sessions were conducted as per the SARI curriculum. Children participated in games and exercises while on horseback to build riding skills while simultaneously receiving therapeutic benefits.

MEASURES

A major issue for psychological studies of TR is lack of viable measures for evaluation. A literature search resulted in a "best fit" option for the current study to collect data mainly through questionnaires. All of the questionnaires used in this study were developed initially by Henderson et al. (2006) and are copyrighted to the ACA (© American Camping Association, Inc., all rights reserved, used by permission). These were scientifically designed and validated surveys that have been used in a variety of camp settings for research purposes (e.g., Henderson et al., 2006; Thurber, Scanlin, Schueler, & Henderson, 2007). The surveys were designed to assess camp experience by campers and parents and the materials were thus acceptable for use with the TR summer camp in the original format. Adaptations to surveys were made for use with the weekly TR sessions; e.g., by replacing the word "camp" with the words "riding-session" and "SARI."

Parents completed an original version of the 'Camper Growth Index – Parent Survey' (CGI-P; acacamps.org), answering pre-camp/course version and a post-camp/course version. Depending on the child's cognitive ability, those attending summer camp completed the pre-camp and post-camp versions of the 'Camper Growth Index – Camper Survey' (CGI-C). No child was capable of doing so for the TR course riding sessions. Additionally, the researchers (A.C. – camp; S.M. – TR courses) completed pre- and post-camp/course checklists, the ACA Camper Growth Index – Observational Checklist (CGI-O). All measures were scored on 4-point Likert scales ranging from 1 (*strongly disagree*) to 4 (*strongly agree*).

The parent questionnaire assessed four main domains titled: positive identity, social skills, physical skills, and positive values. The domains were further divided into eight constructs, titled: self-esteem, independence, leadership, friendship skills, social comfort, peer relationships, physical skills, and positive values. Constructs omitted as not directly applicable to this study were spirituality and environmental awareness. Self-esteem and independence were measured by statements such as "My child believes s/he is a good person" and "My child believes s/he does just fine without parents around." Leadership was addressed by questions such as "My child gets other kids together for games." Friendship skills, social comfort, and peer relationships were tested with statements such as "My child feels s/he can work out problems with other kids", "My child introduces him/herself to new kids" and "My child gets along with others." Physical skills were measured by questions such as "My child likes to try new activities", and positive values were addressed with statements such as "My child helps other people."

The quantitative questions of the parent pre-, post-, and final-camp/session surveys were scored using four-point Likert scales ranging from 1 (*strongly disagree*) to 4 (*strongly agree*). The surveys had 43 quantitative questions, with a possible total score of 172. Questions 12, 14, 21, 28, 31, and 42 of each survey were reverse scored. Mean scores out of a possible maximum of four were reported. The observational checklists had a possible total score of 44. For longitudinal comparisons in study 2, observational checklist percentage scores (from obtained score divided by total score) were calculated.

The researcher observational checklist addressed the same four domains as the parent questionnaires. Positive identity and social skills were measured with statements such as "Is at ease around others" and "Lets others join in activities." Positive values and physical skills were measured with items such as, "Shares/offers to help" and "Shows improved understanding of certain activities."

The camper questionnaires required children to rate their own level of self-esteem, leadership skills, friendship skills, and positive values while at home or at school and while at camp. Questions included, "At camp, I am good at a lot of things" and "Away from camp (at home or at school), I am good at a lot of things".

A 30-minute, semi-structured interview, which had been developed for the camp study, was used to collect qualitative data from the parents for study 1. The interview consisted of open-ended questions that were paired with specific probes in order to clarify what was said, to keep the interview on track, and to obtain the most relevant information possible. The questions addressed the parents' views on the summer camp and TR in general. Each interview was recorded using an Olympus, WS-331 M digital voice recorder in order to efficiently record answers and to obtain a detailed transcript for further analysis. The data collected was categorized by the interviewer (A.C.) by grouping the most common types of responses.

PROCEDURE

Study 1. The pre-camp questionnaires (parent and child) were mailed to the parents and children one week before the child attended summer camp. The questionnaires were mailed with a letter from the executive director at SARI, a letter of information, and all consent (parent) and assent (child) forms. Parents were asked to return the completed surveys on the child's first day of camp. A researcher was present on the first day of camp to answer any questions. One week after the camp was complete hard copies of the post-camp surveys (parent and child) were sent out to all who had completed the pre-camp survey. In addition, a version of the parent and child post-camp questionnaire was made available online via the internet survey company Survey Monkey. A letter explaining the choice for completion of the post-camp surveys was included in the post-camp survey package; this letter contained the URL for the online versions. A stamped return envelope was included with the post-camp survey for parents and children who chose to complete the paper-based survey.

Observational checklists were completed during the camp week. On the first day of camp, the children who had given assent and for whom parental consent had been obtained were observed during the morning activities and riding lessons and the pre-camp checklist was completed. On the last day of camp, these children were again observed by the researcher (A.C.) during the morning activities and riding sessions and the post-camp checklist was completed.

Parents were interviewed one month after the end of the camp session. The interviews were conducted either at SARI, Brescia University College, or the parents' home. Before the interview began, parents were ensured of confidentiality, were told that they would be recorded, and were reminded of their rights as participants.

Study 2. Materials used were as per study 1, with changes to wording (as noted, above) to reflect the TR course use. Parents were given a research package one week before the riding course began.

Riders who had given assent and for whom parental consent had been obtained were observed at the second session of the first 10-week TR course and observational checklists were completed. In this more longitudinal study, riders were observed again at the first and last riding sessions of their second course. At the first and last riding sessions of the second riding course parents completed a post-first course and final-survey at the end of the second TR course; when finished, parents returned the surveys to the researcher. There were no post-course interviews with parents in this study.

For both studies all information collected through questionnaires, checklists, and interviews was coded to maintain confidentiality.

RESULTS

STUDY 1: CAMP PROGRAM

Of the parents initially contacted, 28 completed both the pre-camp questionnaire and the post-camp questionnaire. Of this sample, 20 children had intellectual disabilities, three children had physical disabilities, and five children were of typical ability. An observational checklist was also completed for these 28 campers.

Of the 17 campers who initially completed the pre-camp questionnaire only 14 campers, ranging in age from seven years old to sixteen years old (mean age = 11.1), completed the post-camp survey. The three campers who did not complete the post-camp questionnaire had great difficulty in completing the pre-camp surveys (due to intellectual disability) and the parents felt that his/her answers were unreliable. Data from the 14 campers who completed both pre- and post-camp questionnaires were included. The data from the 28 parents, the 28 checklists, and the 14 campers were scored and subsequently analyzed using paired *t*-tests.

The items on the parent questionnaires and observational checklists were divided into four domains: positive identity, social skills, physical skills, and positive values. Overall, positive camper growth, from the pre-camp period to the post-camp period, was observed in each of the four domains, all p's≤ .01 (see Tables 1 and 2 and Figures 1 and 2).

Table 1: Summary of the means, standard deviations, and t-values for each construct in the parent questionnaires. Note: all comparisons significant to p<.01.

Parent Questionnaires								
Constructs		***Pre-camp***		***Post-camp***				
		M	***SD***	***M***	***SD***	***t-value***	***df***	***p-value***
Positive Identity	*Self-esteem*	2.55	0.64	3.50	0.44	-13.78	27	<.001
	Independence	2.62	0.55	3.27	0.57	-6.96	27	<.001
Social Skills	*Leadership*	2.29	0.68	2.90	0.72	-7.72	27	<.001
	Friendship skills	2.63	0.53	3.35	0.45	-8.85	27	<.001
	Social comfort	2.44	0.50	3.04	0.71	-5.25	27	<.001
	Peer relationships	2.96	0.41	3.41	0.48	-6.95	27	<.001
Physical Skills		2.51	0.79	3.49	0.45	-9.27	27	<.001
Positive Values		2.78	0.59	3.25	0.57	-8.41	27	<.001

Table 2: Summary of the means, standard deviations, and t-values for each construct in the observational checklists. Note: all comparisons significant to p<.01

Observational Checklists							
Construct	***Pre-camp***		***Post-camp***				
	M	***SD***	***M***	***SD***	***t-value***	***df***	***p-value***
Positive Identity	2.04	0.80	2.77	0.91	-6.29	27	<.001
Social Skills	2.17	0.66	2.65	0.83	-4.17	27	<.001
Physical Skills	2.08	0.28	3.11	0.54	-10.13	27	<.001
Positive Values	2.36	0.62	2.68	0.77	-2.78	27	.010

Paired *t*-test comparisons of pre- and post-camp data from questionnaires completed by the children did not reach significance, p = .30.

Parent interviews. From analysis of 15 parent interviews nine constructs emerged that supported the positive camper growth observed: achieving goals, confidence, social comfort, positive values, physical skills, independence, bonding with the animal, communication skills, and normalcy. All of the parents were coded as participant P01, participant P02, and so forth. To preserve confidentiality these codes were used in reporting illustrative quotations from the interviews.

Achieving goals. When asked about goal setting, participant P05 stated: "I think therapeutic riding is so beneficial to special needs kids because they can feel successful while riding. He feels in control, when in life, he is not in control of very much." Participant P09 said: "I've been impressed with my daughter's development and commitment to riding. This has enabled her to reach many different goals."

Confidence. During the interviews, nine parents identified that their child had learnt to be confident and learnt how to trust others. Participant P12 described this increased confidence, saying: "At first she was really scared. But by the end of camp she had learned to trust her horse and trust the volunteers too, a big step for her. By mastering skills the kids gain confidence in themselves". Participant P13 stated, "It really helps his confidence and independence. And it's my opportunity to let him".

Social Comfort. Several parents stated that his/her child enjoyed camp because it was a comfortable environment. Participant P01 explained, "She is confident and at home with the environment there – while riding, with other riders, and with instructors. She is at home and happy."

Positive Values. Other parents explained how camp helped to teach children positive values: "Through the instructor, and the horse, it has taught him compassion, how to love somebody, how to stroke it not poke it, and that sort of thing" (Participant P02). "Camp gave her a chance to be more able than some other kids – it showed her that she can do more things for others" (Participant P07).

Physical skills. The TR at summer camp was identified as being beneficial for children of many different abilities: "Her balance and core strength is improved greatly" (Participant P06). "It has been a positive experience for both developing her gross motor skills and fine motor skills" (Participant P10).

Independence. Participating in the summer camp provides children with a sense of independence. As participant P02 explained, "He doesn't want me to watch him. I'm not his mother when he rides! He says, 'Don't look at me when I'm in there!' He loves the fact that it's something that he can do by himself." while participant P02 stated, "It helps her confidence. It is an activity that we don't have to help, and she feels independent."

Bonding with the animal. The bond that the children seem to form with the horse was the most discussed construct during the interviews. Most parents felt that their children formed special relationships with the animal, as illustrated by these quotes: "Our son loves animals. He forms a special bond with them and is very comfortable around them. He talked about the horse constantly" (Participant P04). "He really loves animals and this gives him an outlet for his affection. I am always happy to see the connection with him and his horse" (Participant P05).

Communication skills. Many parents stated that having to communicate with staff, volunteers, and the horses helped children develop communication skills. Participant P02 explained, "His communication skills have improved immensely, he has a hard time talking properly to people, like he talks 'at' not 'to' and I've seen a lot of improvement in his interactions with other people, yes that has really changed. And I'm giving all that credit to SARI, not to anything else".

Normalcy. The second most discussed construct was normalcy. Parents stated that summer camp was a place for their children to be regarded as normal and was a setting that provided equal standards "SARI is a place where she can be herself and not worry about being different or the pressures of fitting in. She is not alienated like she is in the school environment" (Participant P01). Overall, parents rated the SARI summer camp as being a very beneficial experience. As stated by participant P02, "This is the best way to teach kids to: show love, respect others, show kindness, meet other people, no judgment, higher self-esteem".

STUDY 2: WEEKLY SESSIONS

Participants for study 2 ranged in ages from 5-18 years. Of the 18 riders and parents initially contacted to participate in this study, only nine riders (mean age = 9.4) completed two, 10-week TR courses. Moreover, only nine parents completed the three pre-, post- and final-course surveys. Although it was intended for children to complete pre-, post-, and final-child riding-session surveys, no children completed these surveys due to a lack of capability. Data from nine researcher observational checklists and the nine parent riding-session surveys thus were used in the statistical analyses performed for this study.

The questionnaire and observation scores were calculated as before and analyzed for this study using a one-way analysis of variance (ANOVA) to compare the mean percentage scores (scores calculated as percentage of possible maximum) across three separate time points: (1) pre-TR course; (2) post-first course (after the first TR course, at the start of the second); and (3) final, at the end of the second TR course.

Comparison of scores for the parent TR course surveys showed no effect; $F(2, 24) = 2.63$, *ns*. Figure 3 shows that there was a numerical, though not statistically significant, increase in scores.

A significant positive effect was found, however, for the mean percentage scores for the researcher observational checklists across the three time points (pre-course one, post-course one, and final, at end of TR course two), $F(2, 24) = 39.21$, $p < .01$ (see Figure 4). More specifically, a significant increase in the mean percentage scores was found between time-point one and time-point two, and between time-point one and time-point three. Post-hoc t-tests determined that there was a significant increase in the percentage scores of the researcher observational checklists between time-point one ($M = 58.90$, $SD = 5.96$) and time-point two ($M = 85.10$, $SD = 8.38$), $p < .01$, and also between time-point one ($M = 58.90$, $SD = 5.96$) and time-point three ($M = 91.03$, $SD = 9.78$), $p < .01$. However, there was no significant difference in percentage scores between time-point two ($M = 85.10$, $SD = 8.38$) and time-point three ($M = 91.03$, $SD = 9.78$), ns.

Discussion

The results supported the hypothesis that children would experience an increase in positive identity, social skills, physical skills, and positive values from pre- to post-TR experience. Overall, the results from the parent questionnaires and observational checklists suggest that through engagement in TR, especially at summer camp, children experienced increased confidence, learnt to believe in their abilities, and were more comfortable in performing activities without parental assistance.

Study 1: Camp Experience

In addition to the effects noted above, campers improved at taking the lead in activities and at playing and sharing with other campers. Children also learnt how to be comfortable in a new environment, were more likely to try new activities throughout the week of camp, and developed new skills. The campers experienced growth in positive values such as respecting others, helping others, and following rules. Constructs derived from parent interviews indicated that parents saw a positive growth in their children in the following areas: achieving goals, confidence, social comfort, positive values, physical skills, independence, bonding with the animal, communication skills, and normalcy.

These results are consistent with previous research findings that camps are beneficial for all children (Henderson et al., 2007; Thurber et al., 2007). The results are also consistent with studies stating that camps are beneficial to children with disabilities. Children with disabilities who participate in camps show significant improvements in behaviour, attitude, and social skills (Michalski et al., 2003; Shaver & Scheibe 1967; Van Belois & Mitchell, 2009).

To the authors knowledge there are no published studies on effects of TR in a camp setting on special needs populations. Most research investigating TR has been on physical disabilities

(Snider et al., 2007; Lehrman & Ross, 2001; Davis et al., 2009) or on adults with disabilities (Karol, 2007; Corring et al., n.d). Karol (2007) found that interaction with horses helped adults with disabilities develop responsibility, autonomy, awareness, and communication skills, and the present study found TR to affect children with disabilities in a similar fashion. It was also discovered from our studies that TR appeared to be especially beneficial at camp, probably because it was intensive (i.e., the children rode every day). Therefore, children of all abilities were able to learn and improve much more quickly than they would only on a weekly basis.

Study 2: Weekly Session TR

Children at regular TR sessions experienced improvements in four developmental domains: positive identity, social skills, positive values, and physical skills. Riders' percentage scores on the observational checklists increased significantly from time-point one to time-point two, and from time-point one to time-point three; however, children did not show significant positive effects from TR between time-point two and time-point three. It can be concluded then that these children experienced significant beneficial differences from TR from the beginning of TR sessions to the end of TR riding sessions, but that any benefits were gained more slowly over the longer term with weekly TR sessions.

The results from the observational checklists indicated that children showed significant improvements in self-esteem and, as with summer camp TR, they were more willing to try new activities. Riders showed improvements in independence, and were more willing to volunteer to go first and to ride independently, without assistance from a side-walker, leader, or instructor. Additionally, a significant increase in riders' social comfort was observed (i.e., with other riders, side-walkers, leaders, and instructors); moreover, children's social skills improved as they were more likely to let other riders join in activities and were able to make friends more comfortably. Finally, it was observed that children's physical skills significantly improved as the children in general showed better coordination and significant improvement in the TR activities.

Although results from the parent riding-session surveys did not reach significance, a numerical increase in the percentage scores of parent riding-session surveys was found. Qualitative data ("additional comments" sections on the parent surveys) was also collected and confirmed that parents noticed positive changes in their children. In general parents stated they saw overall improvements in their child's ability and willingness to do the following activities on a consistent basis: speak more clearly, show patience, talk to new people, and follow directions more frequently. Parents also saw a substantial increase in their child's confidence and in their balance. Nearly all of these qualitative changes noted by parents are similar to results of previous research such as that conducted by Dranch et al. (2010), Davis et al. (2009), and Bass et al. (2009). Specific examples from parent survey comments also illustrated the improvements noted. Thus, even though parent riding session surveys did not yield significant changes across time points, parent qualitative responses did note perceived positive differences in their children after two 10-week riding sessions.

General Issues

Despite the overall positive results, there were limitations to these studies. First, the pre-camp and post-camp surveys were only investigating the short-term effects of camp and did not determine whether there are any positive long-term effects on the children. Although in the regular sessions children were tracked across three time points, no follow-up interviews were obtained. The lack of long-term follow-up is an oversight which should be dealt with in future studies of this type. Second, since the sample was composed of mostly children with disabilities, small changes could lead to exaggerated results of general effects of riding in this setting. As stated by Thurber et al. (2007), children who scored low on pre-camp surveys had the most "room for improvement" in post-camp surveys. The majority of the children in our camp study scored fairly low on the pre-camp parent questionnaires and observational checklists and experienced a large positive growth in post-camp measures. Due to the nature of most of the disabilities, a small change in a child's behavior could make a large impact on his/her parents and the researcher. This means that campers may not have experienced as large an improvement as might appear. Another limitation to our study may be that neither the researchers nor the participants were blinded to the study interventions in order to reduce the potential for bias. Parents would have viewed TR as a potentially positive activity to have enrolled their child in the first place. The length of time between pre- and post-camp/course surveys would have mitigated biasing effects, however, and efforts were made by the researchers to be as objective as possible. It should be noted also that neither parents nor researchers had the original assessments (surveys or observation records) to refer to when filling out post-camp/course documents. Finally, participant numbers were a significant issue. In the TR course study (Study 2) only nine parents and their children participated fully. For both studies sample sizes were smaller than had been originally anticipated due to children tending to attend multiple TR sessions; e.g., some campers attended more than one week of camp.

Because of the lack of research on the effects of TR generally, and at summer camp particularly, there is much potential for further studies. It would be ideal, as previously stated, to investigate the long-term effects of TR; e.g., by using follow-up surveys of parents and riders four or six months after the completion of TR experience. This would provide valuable insight regarding the long-term benefits of TR, especially for the one-time experience of the summer camp. It may also be worthwhile to gather data from TR institution staff since, as previously mentioned, children tend to overestimate their skill level, adjustment, and competence (Klassen & Lynch, 2007; Thurber et al., 2007) and gathering data from different, reliable sources would provide a more objective evaluation (Henderson et al., 2007; Klassen & Lynch, 2007; Michalski et al., 2003; Thurber et al., 2007). Additionally, as not all of the children in the current studies (none in Study 2) completed child riding-session surveys, the adaptation of a child riding-session survey would likely prove beneficial. The design of parent riding-session survey could be altered to questions only regarding children's behavior; a simpler survey would make completion of the survey easier for parents, and would likely yield more consistent answers.

Finally, the four main domains tested encompassed a large number of constructs; therefore, it may be useful to test the constructs separately and more intensively in order to determine which ones provide the most positive and lasting effect and if different settings or circumstances affect these certain constructs. A larger sample size would allow for additional comparisons to be made between and within participants. Greater participant numbers would enable comparisons, for example, by age group or by specific condition (e.g., cerebral palsy or autism spectrum disorder).

Conclusions

Centres such as SARI provide specialized support and adapted programs to children with special needs. Parents, staff, and health professionals who are involved with TR believe that this activity provides rehabilitation and cognitive, psychological and physical development for the children. It is also said that through TR children can experience greater freedom and independence. For many participants, TR is one of the only activities that they can experience. TR is truly a unique activity that provides skills for children with special needs that may be transferred into their everyday lives. It is especially beneficial to experience TR in a camp setting as it tends to be an intensive setting and children, parents, and staff notice positive changes much more rapidly. From the results obtained from the multiple measures in this study, it would appear that TR has beneficial effects for children with special needs' social functioning, physical functioning, and positive self-perceptions. Thus TR, especially in an intensive summer camp setting, may provide children with special needs the social and physical experience needed to effectively develop. Thus TR could be recommended to any child with special needs seeking social interaction with others and regular physical exercise.

There is, unfortunately, a lack of options and a lack of funding for programs that accept special needs populations (Brookman et al., 2003; Michalski et al., 2003). Therefore, it is important that advances in research are made to create awareness and, in turn, create more TR opportunities, opportunities that foster positive growth.

REFERENCES

American Camp Association (n.d.) *Camper Growth Index (CGI) Surveys*. Retrieved from http://www.acacamps.org/research/enhance/directions/CGI

American Youth Horse Council & Pennsylvania State University (2006). *Impact of equine activities on youth development*. Colorado Springs, CO: American Youth Horse Council.

Bass, M. M., Duchowny, C. A., & Llabre, M. M. (2009).The effect of therapeutic horseback riding on social functioning in children with autism. *Autism Development Disorder*, *39*, 1262-1267.

Beck, A. M. (2011). Animals and child health and development. In: P. McCardle, S. McCune, J. A. Griffin, L. Esposito, & L. S. Freund (Eds.), *Animals in our lives: Human-animal interaction in family, community and therapeutic settings* (pp. 43-52). Baltimore, MD: Paul H. Bookes Publishing Co..

Bizub, A. L., Joy, A., & Davidson, L. (2003) "It's like being in another world": Demonstrating the benefits of therapeutic horseback riding for individuals with psychiatric disability. *Psychiatric Rehabilitation Journal, 26*, 377-384.

Brookman, L., Boettcher, M., Klein, E., Openden, D., Koegel, R. L., & Koegel, L. K. (2003). Facilitating social interactions in a community summer camp setting for children with autism. *Journal of Positive Behaviour Interventions, 5*(4), 249-252.

Burt, K.B., Obradovic, J., Long, J.D., & Masten, A.S. (2000). The interplay of social competence and psychopathology over 20 years: Testing transactional and cascade models. *Child Development, 71*(4), 1049-1060.

CanTRA (n.d.) About CanTRA. Retrieved from http://www.cantra.ca/aboutCantra.shtml

CanTRA (n.d.) CTRII Certification Information. Retrieved from http://www.cantra.ca/CTRII.shtml

Catalano, R. F., Berglund, M. L., Ryan, J. A. M., Lonczak, H. S., & Hawkins, J. D. (2004). Positive youth development in the United States: Research findings on evaluations of positive youth development programs. *Annals of the American Academy of Political and Social Science, 591*, 98-124.

Colyn, L., DeGraaf, D., & Certan, D. (2008, March/April). Social capital and organized camping. It's about community! *The Camping Magazine, 81*(2), n.p.. Retrieved from http://www.acacamps.org/members/knowledge/participant/cm/0803socialcapital

Corring, D. J., Johnston, M., & Rudnick, A. (n.d.) *The effects of a supported program for horse-back riding in inpatients diagnosed with Schizophrenia: A qualitative Exploratory Study.* Unpublished report from University of Western Ontario and Regional Mental Health Care to SARI.

Davis, E., Davies, B., Wolfe, R., Raadsveld, R., Heine, B., Thomason, P., Dobson, F. & Graham, H. K. (2009). A randomized controlled trial of the impact of therapeutic horse riding on the quality of life, health, and function of children with cerebral palsy. *Developmental Medicine & Child Neurology, 51*, 111-119.

DePauw, K. P. (1986). Horseback riding for individuals with disabilities: Programs, philosophy, and research. *Adapted Physical Activity Quarterly, 3*, 217-226.

Dranch, M., O'Brien, P. A., & Kreger, A. (2010). The effect of a 5-week therapeutic horseback riding program on gross motor function in a child with cerebral palsy: A case study. *The Journal of Alternative and Complementary Medicine, 16,*(9); 1003-1006.

Dworken, B.S. (2001). Research reveals the assets of camp. *The Camping Magazine, 74*(5), 40-43.

Esposito, L., McCardle, P., Maholmes, V., McCune, S., & Griffin, J. A. (2011). Introduction. In: P. McCardle, S. McCune, J. A. Griffin, L. Esposito, & L. S. Freund (Eds.), *Animals in our lives: Human-animal interaction in family, community and therapeutic settings* (pp. 3-5). Baltimore, MD: Paul H. Brookes Publishing Co..

Gaskell, S.L. (2007). The challenge of evaluating rehabilitative activity holidays for burn-injured children: Qualitative and quantitative outcome data from a burns camp over a five-year period. *Developmental Neurorehabilitation, 10*(2), 149-160.

Gendron, M., Royer, E., Bertrand, R., & Potvin, P. (2004). Behaviour disorders, social competence and the practice of physical activities among adolescents. *Emotional and Behavioural Difficulties, 9*(4), 249-259.

Henderson, K. A., Schueler Whitaker, L., Bialeschki, M. D., Scanlin, M. M., & Thurber, C. A. (2007). Summer camp experiences: Parental perceptions of youth development outcomes. *Journal of Family Issues, 28*(8), 987-1007.

Henderson, K. A., Thurber, C. A., Schueler Whitaker, L., Bialeschki, M. D., &Scanlin, M. M.(2006). Development and application of a camper growth index for youth. *The Journal of Experiential Education, 29*(1), 1-17.

Karol, J. (2007). Applying a traditional individual psychotherapy model to equine-facilitated psychotherapy (EFP): Theory and method. *Clinical Child Psychology and Psychiatry, 12*(1), 77-90.

Klassen, R. M., & Lynch, S. (2007). Self-efficacy from the perspective of adolescents with LD and their specialist teachers. *Journal of Learning Disabilities, 40*(6), 494-507.

Lehrman, L. & Ross, D. B. (2001). Therapeutic riding for a student with multiple disabilities and visual impairment: A case study. *Journal of Visual Impairment & Blindness, 93*(6), 108-109.

Michalski, J. H., Mishna, F., Worthington, C., & Cummings, R. (2003). A multi-method impact evaluation of a therapeutic summer camp program. *Child and Adolescent Social Work Journal, 20*(1), 53-76.

Rinn, A. N. (2006). Effects of a summer program on the social self-concepts of gifted adolescents. *The Journal of Secondary Gifted Education, 17*(2), 65-75.

Shaver, P.R., & Scheibe, K.E. (1967). Transformation of social identity: A study of chronic mental patients and college volunteers in a summer camp setting. *Journal of Psychology, 66*(1), 19-37.

Siperstein, G. N., Glick, G.C., Bardon, J. N., Harada, C. M., & Parker, R. C. (2007, Jul/Aug). Camp Shriver: A Model for including children with intellectual disabilities in summer camp. *The Camping Magazine*, *80*(4), 20-26.

Snider, L., Korner-Bitensky, N., Kammann, C., Warner, S., & Saleh, M. (2007). Horseback riding as therapy for children with cerebral palsy: is there evidence of its effectiveness? *Physical & Occupational Therapy in Pediatrics, 27*(2), 5-23.

Sterba, J. (2007). Does horseback riding therapy or therapist-directed hippotherapy rehabilitate children with cerebral palsy? *Developmental Medicine & Child Neurology*, 49; 68-73.

Thurber, C. A., Scanlin, M. M., Scheuler, L., & Henderson, K. A. (2007). Youth development outcomes of the camp experience: Evidence for multidimensional growth. *Journal of Youth Adolescence, 36*, 241-254.

Van Belois, A. & Mitchell, J. B. (2009, Nov/Dec). Summer residential camps: Promoting social interactions and self-efficacy among young adults with special needs. *The Camping Magazine, 82*(6), 40-43.

Wells, D. L. (2009). The effects of animals on human health and well-being. *Journal of Social Issues, 65*(3), 523-543.

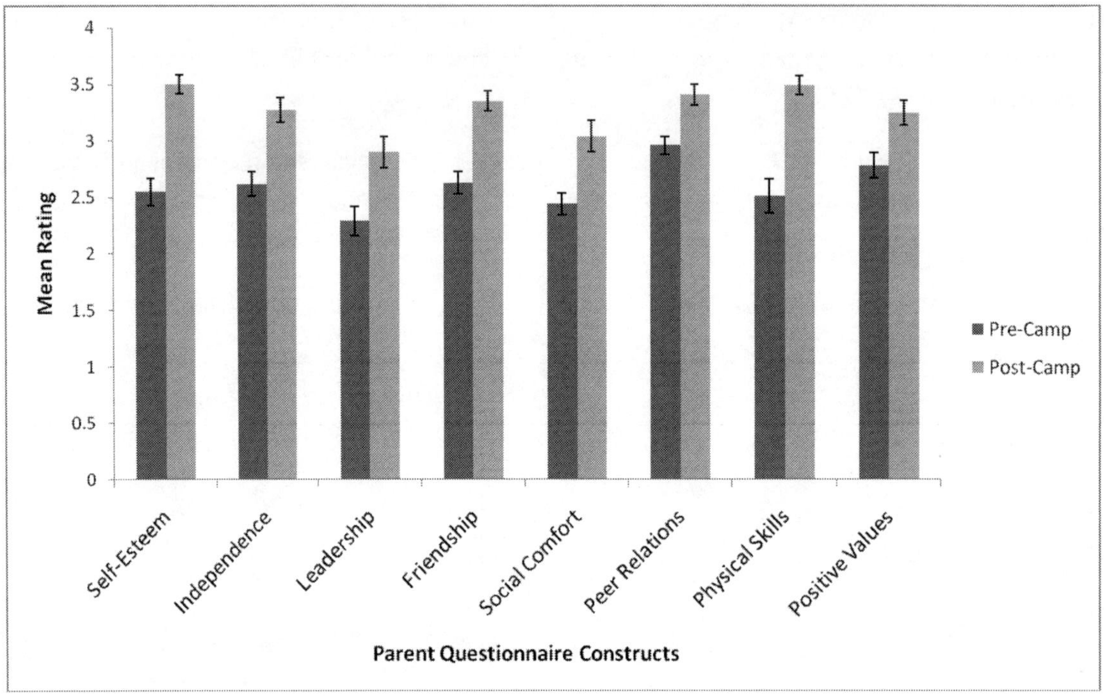

Figure 1. Combined mean parent ratings for their child across the eight constructs. A significant growth across the camp period was observed in all of the constructs, *p*< .01.

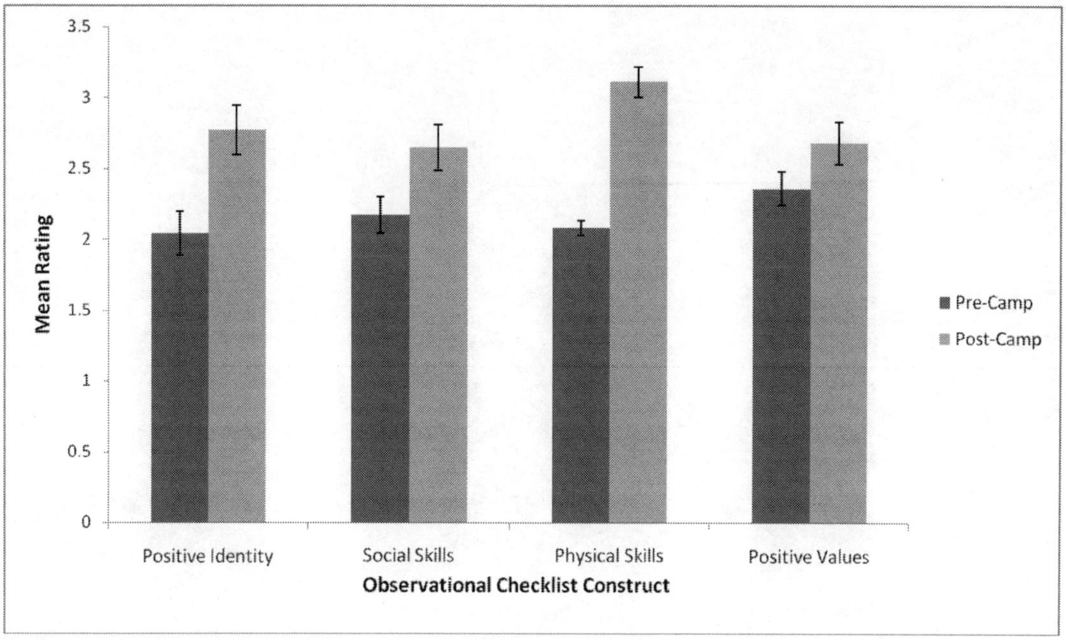

Figure 2. Mean ratings across campers for the four domains on the observational checklist. A significant difference from the pre-camp to post-camp period was reached in all four domains, $p < .01$.

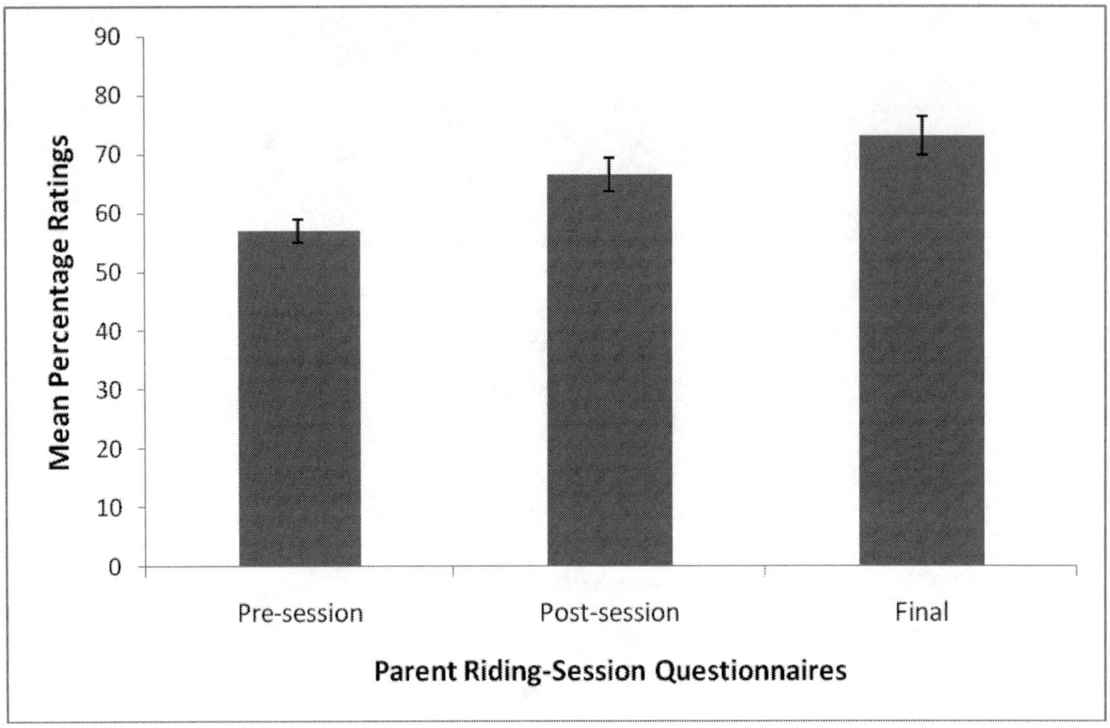

Figure 3. Combined mean percentage ratings by parents for their child from the pre-session to the final questionnaire. All comparisons *ns*.

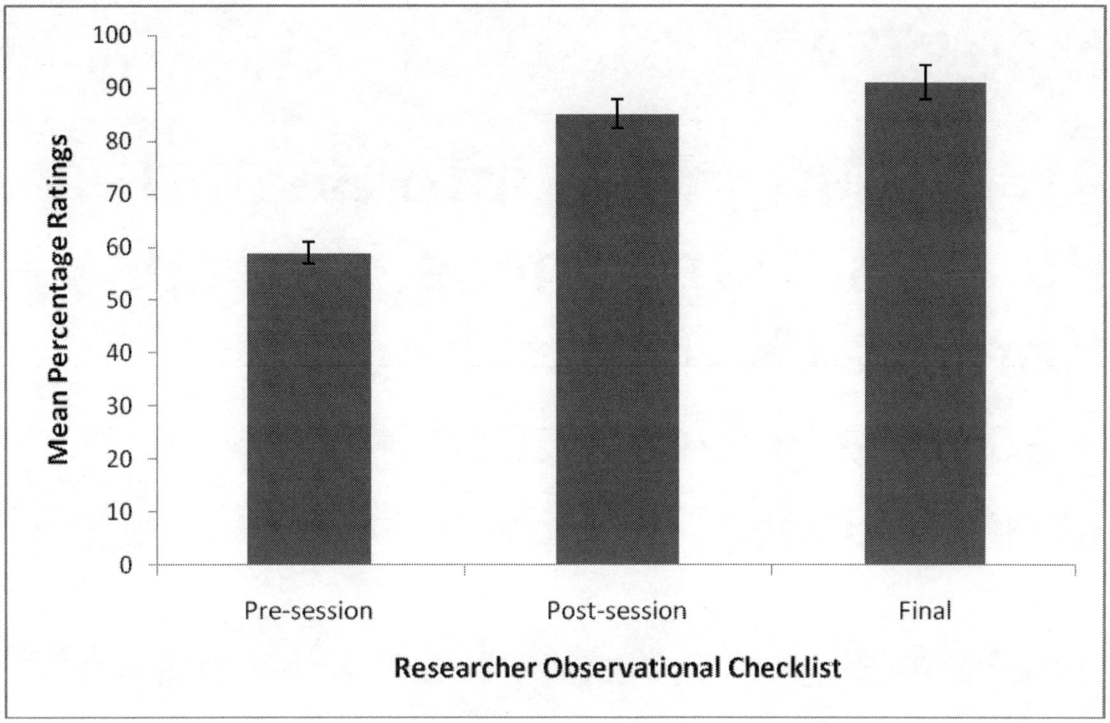

Figure 4. Mean percentage ratings for campers across riding session on the observational checklist. A significant difference was found between pre-session and post-session ($p<.01$), and pre-session and final ($p<.01$).

ARTICLE TWO

An Investigation into Seated Postural Differences between Horseback Riders and Sedentary Non-riding Individuals

Patrycja Czarnoleska

Correspondence to:
Patrycja Czarnoleska
19 Tamhnach Li,
Monavalley, Tralee,
Co. Kerry Ireland
Email: pattyme6@hotmail.com

ABSTRACT

Postural problems are prevalent due to lifestyle factors. Horseback riding might have the potential to improve the postural configuration of individuals, but evidence is sparse. The aim of this study was to investigate the postural differences between horseback riders and sedentary non-riding individuals. In this small-scale study, quantitative and qualitative measurements were used to assess sagital plane, seated posture of a group of 43 individuals (25 females, 18 males) including 23 horseback riders and 20 non-riders. There were no statistical anthropometrical, age, gender or personality differences between riders and non-riders and the confounding factors of psychosocial and socio-economic profile were eliminated. Riders showed significantly better posture in some variables, namely cranio-cervical angle, neck flexion angle, position of shoulder and pelvic tilt. Thus, there is preliminary evidence that horse riding may aid human posture.

KEYWORDS

Horseback riding, posture

INTRODUCTION

Good posture is defined as a: "...state of muscular and skeletal balance which protects the supporting structures of the body against injury or progressive deformity, irrespective of the attitude (erect, lying, squatting or stooping) in which these structures are working or resting" (Kendall, McCreary, Provance, Abeloff, 1993, p. 4). Proper posture (appropriate bone-muscle-connective tissue relationship) allows for optimal functioning of the body. Correct posture: helps with energy efficiency and even distribution of blood and nutrients to vital organs of the body; allows for optimum use of vision and hearing through the positioning of the head; reduces muscle strain and provides an optimal attachment for the various ligaments in the human body; facilitates functioning of the heart and attached viscera (Johnson, 1998; Pulig-Schatz, 1992).

More and more people are experiencing postural issues (Darling, Kraus, & Glasheen-Wray, 1984; Harrison, Barry-Greb, & Wojtowicz, 1996; Johnson, 1998; Lewis, Wright, & Green, 2005; McDonnell, 2006; MacEvoy & Grimmer, 2005; Yoo, Cho, Jeon, Cynn, & Choi, 2008). This is often linked to lifestyle factors. People are spending much of their working lives as well as their free time in front of the television or computer. The increased time in a sitting position imposes stresses on the human body and may result in a faulty posture, discomfort or even pain (Kendall et al., 1993). Moreover, increased prevalence of postural problems causes higher healthcare costs (Ferrari & Russell, 2003).

Various methods have been used to enhance posture such as Pilates, swimming or Tai-Chi. Therapeutic horseback riding is one such method with a long history and some support for its value (Bertoti, 1988). One of the basic skills in horseback riding is to "stay in a correct balanced position whilst absorbing the movement of the horse" (Kyrklund & Lemkow, 1998). In technically correct horseback riding posture, the head and neck need to be well balanced. Horseback riding can improve posture, equilibrium reactions, sitting balance, standing balance and even walking (MacKinnon, Noh, Laliberte, Allan, & Lariviere, 1995).

There are many studies advocating multiple benefits of horseback riding such as improving balance, co-ordination and postural control (Bertoti, 1988; Debuse, Chandler, & Gibb, 2005; Gough, 1993; Haehl, Giuliani, & Lewis, 1999; Heipertz, 1981; Hornacek, 2005; Land, Errington-Povalac, & Paul, 2001; MacGibbon, Andrade, Widener, & Cintas, 1998; MacKinnon & Ferreira, 2002; Wham & Cert, 1990; Would, 2002). Additional positive outcomes include: greater lung capacity, improved motor performance, increased stamina, enhanced symmetrical weight bearing and muscle activity, better head position control, correction of abnormal movement patterns, improved gait, regulation of muscle tone, strengthening of trunk stabilisation muscles and improvement in proprioception and spatial awareness (Debuse et al., 2005; Harrington, 1999; Hornacek, 2005; Ionatamishvilli, Tsveravaq, & Abaliani, 2001; Sakakura, de Barros Santos, R., Silkwood-Sherer & Warmbier, 2007; Cyrillo, Perdigao, & Torriani, 2006; Would, 1996). Therapeutic riding is also claimed to benefit spastic and hypotonic conditions (Brock, 1989).

Horseback riding mimics three-dimensional rhythmical movement of human pelvis during normal gait, mobilizes the pelvis, improves alignment and trunk control, enhances voluntary strength of abdominal, trunk extensor, hip extensor and shoulder musculature and also sitting balance (Bertoti, 1988; Gough, 1993; Land et al, 2001). The rider needs to adjust to the pelvic movement of a horse and learns to “discover” proper solutions using innate balancing capabilities (Silkwood-Sherer & Warmbier, 2007). The correct riding posture should incorporate features such as upright pelvis, erect trunk and slight lordosis (Would, 1996). The rider has to use its balance reactions and raise the muscle recruitment in order to maintain itself on the moving horse (Sakakura et al., 2006). Each step of the animal challenges a postural response from the rider’s body. This response is practiced over and over again as the horse moves (Wham and Cert, 1990). Horse-riding exercises in a dynamic way the muscles involved in keeping a balanced, erect position. The straight posture and rhythmic movement when horseback riding, facilitates breathing. While sitting on a moving horse the individual exercises deep postural muscles from a corrected position (Gough, 1993: Mann 2005).

Analysis of horse and human walking gaits confirmed that they are corresponding and are therapeutic in nature. Therefore, horseback riding could have the potential to improve musculoskeletal configuration of human posture (Uchiyama, Ohtani, & Ohta, 2011). What is unique is the rhythmical movement which is not a mechanical beat but a continuous repetition of movement stimuli in a dynamic, three-dimensional way. The horse balances the rider with every step and the horse and the rider’s centre of gravity match on a vertical plane (Strauss, 1998). Many skills involved in therapeutic horseback riding can be used to a greater degree by any rider (MacKinnon & Ferreira, 2002).

The following investigation was conducted because the existing literature on positive influences of horseback riding on posture was sparse. This particular study differed from the previous investigations in using both qualitative and quantitative posture measurements, sagital plane sitting position mimicking horseback riding and inclusion of psychosocial profile of participants.

The aim of this study was to investigate seated postural differences between a group of recreational horseback riders (riding at least 3 times a week for a period of minimum 5 years) and a group of sedentary non-riders (claiming not to be involved in physical activity on a regular basis).

Method

Participants and Setting

A total of 43 participants (25 males, 18 females), 23 horseback riders (13 females and 10 males) and 20 non-riders (12 males and 8 females) participated in the study. The riders had to be riding at least 3 times a week for a period of minimum 5 years. The 20 non-riders were required not to be involved in physical activity on a regular basis. The horseback riders were selected from The Kerry Riding Club and Fairies Stables in county Kerry. The non-riders were staff members of the Institute of Technology, Tralee, Ireland.

Measures and Procedure

The measures used in this study comprised:

- Questionnaire designed to gather anthropometric and psychosocial data – including social class and personality profile (based on the work of Nettle, 2007)
- Quantitative analysis of posture (angles from a photograph)
- Qualitative postural analysis using Bertoti's Posture Assessment Scale (observation)

The methodology was pilot tested, resulting in very minor modification of the research process. Consent forms were obtained before the commencement of the study.

Participants' height and weight measures were taken and their handedness and psychosocial profile established (through the use of a questionnaire).

During a scientific photography session markers were placed on the side of the dominant hand using palpation methods. Sites where markers were placed included (see Figure 1):

(A) the tragus of the ear,
(B) the cantus of the eye,
(C) C7 on the back of the neck,
(D) the acromion process on the shoulder,
(E) the greater trochanter on the side of the hip
(F) ASIS (anterior superior iliac spine).

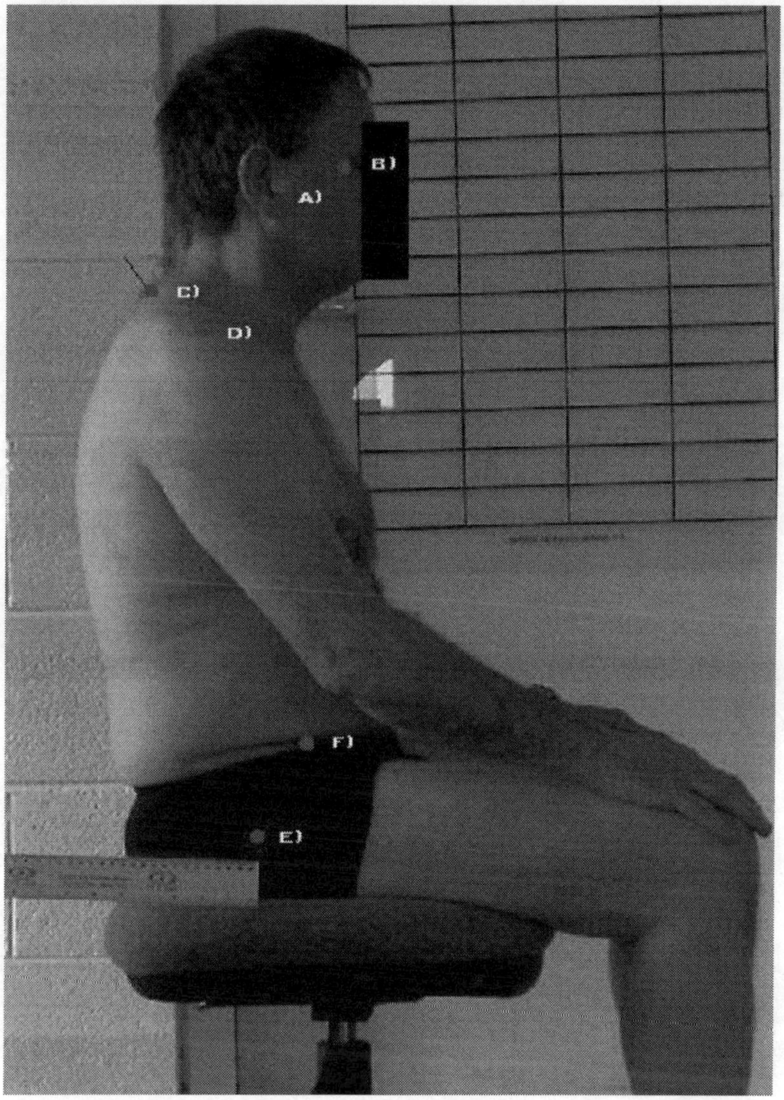

Note: A) the tragus of the ear; B) the cantus of the eye; C) C7 on the back of the neck; D) the acromion process on the shoulder; E) the greater trochanter on the side of the hip; F) ASIS (anterior superior iliac spine).

Figure 1. Body landmarks used in the study

Then, participants were asked to sit on the height-adjustable chair and the angle at the knee was checked to be 90 degrees. They were then instructed to look at a fixed point on the wall and to rock the upper body slightly forward and back to find the most comfortable (and balanced) position. This method was previously used by Watson (1993). Next, two sagittal plane photographs were taken by a qualified photographer and the researcher used the Bertoti's Posture Assessment Scale to make a visual assessment at the same time. Sagittal plane sitting posture was assessed quantitatively through photographic analysis of visual markers placed on bony landmarks. The angles and distances were determined using a digital software called *ImageTool.*

Figure 2 illustrates the postural angles used in this research study. The definitions were as follows:

(1) head flexion (angle measured between the vertical and the line from canthus to tragus – measured above intersect),

(2) gaze angle (between the horizontal and the line from canthus to tragus – measured anterior to intersect),

(3) cranio-cervical angle (between the line from canthus to tagus and tragus to C7 – measured anterior to intersect),

(4) neck flexion (angle between the vertical and the line from tragus to C7 – measured above intersect),

(5) cranio-vertebral angle (between the horizontal and the line from tragus to C7 – measured anterior to intersect),

(6) acromio-vertebral angle (between the horizontal and the line from C7 to acromion process – measured below the intersect),

(7) pelvic tilt (angle between the vertical and the line from ASIS to greater trochanter – measured from vertical above the intersect),

(8) shoulder thrust (the distance between the verticals intersecting both C7 and acromion process),

(9) neck angle (between lines from tragus to C7 and C7 and greater trochantor – measured anteriorly to intersection).

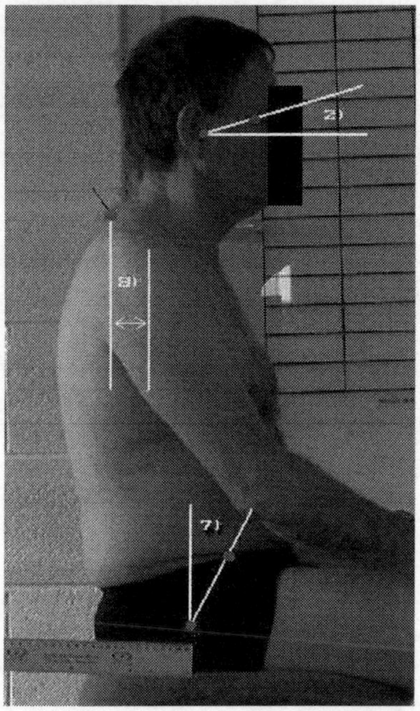

<u>In picture 1:</u>

- gaze angle (2),
- shoulders thrust (8)
- pelvic tilt (7)

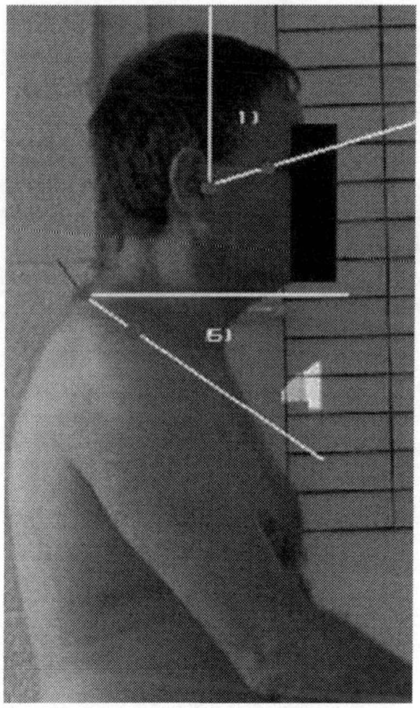

<u>In picture 2:</u>

- head flexion (1)
- acromio-vertebral angle (6)

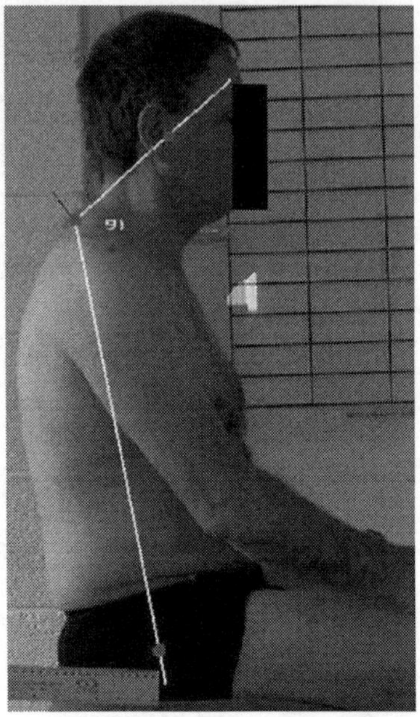

<u>In picture 3:</u>

- neck angle (9)

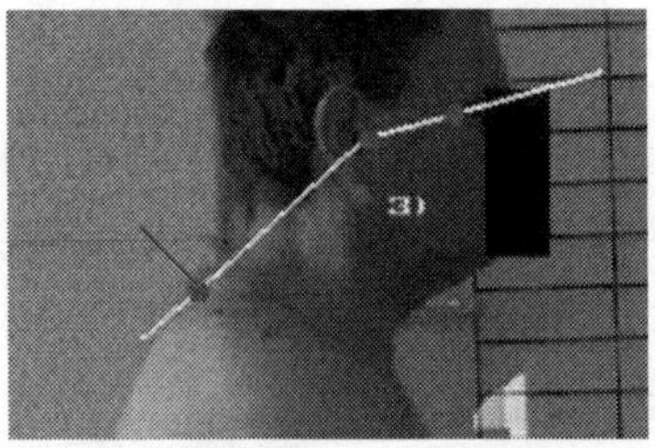

<u>In picture 4:</u>

- cranio-cervical angle (3)

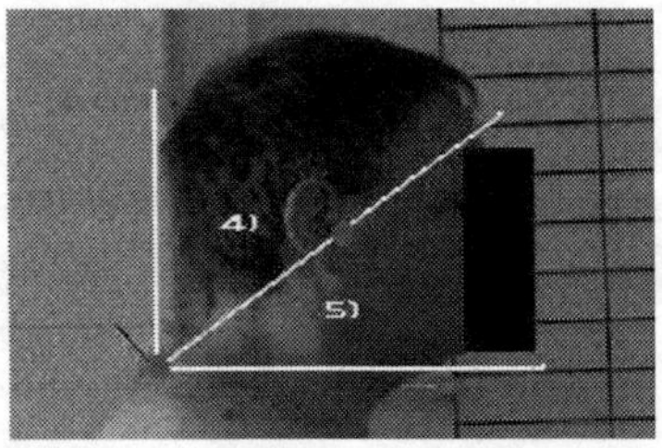

<u>In picture 5:</u>

- neck flexion (4)
- cranio-vertebral angle (5)

Figure 2. Angles measured from digital photographs using *ImageTool* digitising software.

The angles measured in this study have been used in previous research papers. Straker O'Sullivan, Smith, & Perry (2009) utilised head and neck flexion, cranio-cervical angle and pelvic tilt measures. Ankrum & Nemeth (2000) as well as MacEvoy & Grimmer (2005) used gaze angle (also called head tilt). Neck angle was previously used by Burgess-Limerick, Plooy, Fraser, & Ankrum (1999) and cranio-vertebral angle was utilised by Cureton (1941), Darling et al. (1984), Watson (1993), Gillian (1998), Ankrum & Nemeth (2000), Lewis et al. (2005) and Yoo et al. (2008). Yoo et al. (2008) used also acromio-vertebral angle in his study.

Bertoti's Posture Assessment Scale was designed to elicit information about postural configuration focusing on four main areas: head and neck, shoulders, spine and pelvis. The score ranging from 0 to 3 was assigned for postural alignment for each of the areas, where 0 indicated severe misalignment and 3 good, stable and balanced posture. The total score was calculated for each individual by adding the scores for each of four body areas. Higher total scores indicate better overall postural alignment. Originally in the observational scale devised by Bertoti (1988) five trained physical therapists rated subjects anteriorly, laterally and posteriorly by walking around them. In this research study an adapted version of the scale was used and the author utilised only the lateral (side on) view. The scale was used by a trained individual and double checked by the researcher while analysing the digital photographs.

RESULTS

PARTICIPANT CHARACTERISTICS

The majority of horseback riders were between the age of 25 and 30 and also 46 and 50. On the other hand, the majority of non-riders were between the age of 25 and 30 and 36 and 40 (see Figure 3). Despite the variations there was no significant difference in between the groups' age ranges ($p>0.05$).

There was an equal distribution of both genders in this study and no significant difference was found ($p>0.05$).

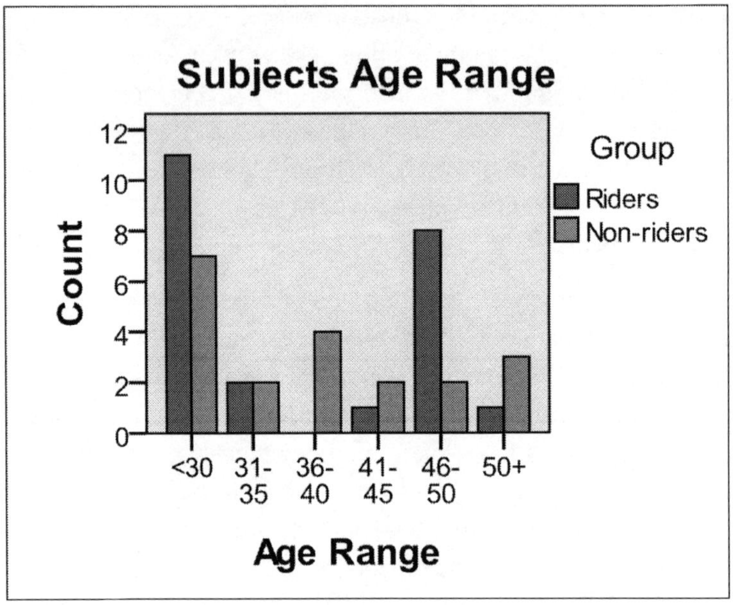

Figure 3. Age distribution for riders and non-riders

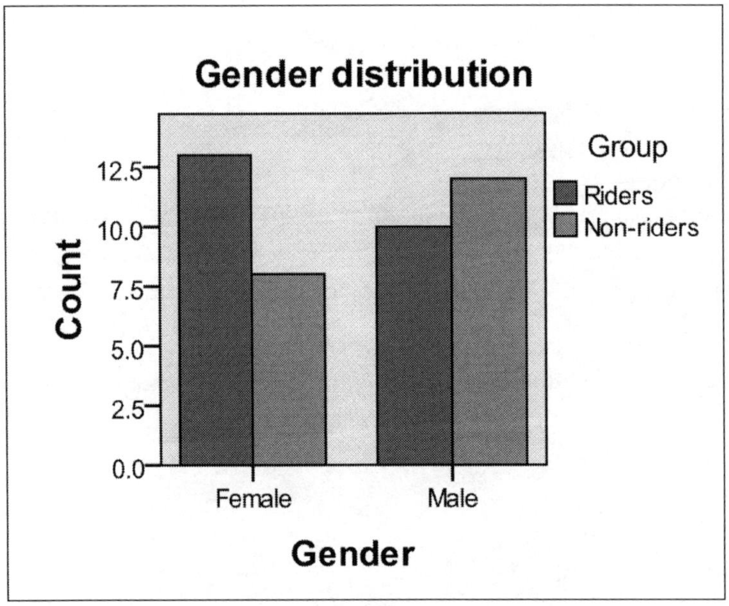

Figure 4. Gender distribution for riders and non-riders

Non-riders tended to be from a higher social class than riders (see Table 1). Most participants from the riding group were skilled workers compared to predominantly white collar members of the non-riding group . Despite this difference there was no significant difference between the groups (*p*>0.05).

The higher the score, the participants were more likely to be presented with a particular personality trait (see Table 2). As previously noted, the riders did not differ significantly from non-riders in their personality profile (p>0.05).

Table 1: Social class systems

SES N=43	Riders	Non-riders
Episodes of employment and unemployment	6	4
Skilled worker	11	4
White collar	4	11
Higher professional	2	1

SES- socio-economic status

Table 2: Personality profile (Nettle, 2007)

N=43	Mean	Std. Dev.
Extraversion (all)	**6.88**	**1.8**
Riders	7.3	1.64
Non-riders	6.4	1.9
Neuroticism (all)	**4.81**	**1.85**
Riders	5.26	1.89
Non-riders	4.3	1.72
Conscientiousness (all)	**6.93**	**1.76**
Riders	6.83	1.61
Non-riders	7.05	1.96
Agreeableness (all)	**12.23**	**2.31**
Riders	12.17	2.71
Non-riders	12.3	1.81
Openness (all)	**8.86**	**2.7**
Riders	8.13	2.72
Non-riders	9.7	2.49

No statistically significant difference found in any of the variables

For the 43 participants, the mean height of all participants was 172.4 cm and weight was 74.46 kg. There was a slight variation between the groups with non-riders being a bit taller and heavier (see Table 3).There was no gender differences. The mean body mass index (BMI) score of all subjects was 25 where non-riders scored 26.2 and riders 24. There were no between group significant differences found for any of the anthropometric variables between riders and non-riders groups (all $p>0.05$).

Table 3:- Subjects' Anthropometric Data

N=43	Mean	Std. Dev.
Height (all) (cm)	**172.4**	**6.87**
Riders	171.58	8
Non-riders	173.35	5.29
Weight (all) (cm)	**74.46**	**14.70**
Riders	70.91	13.06
Non-riders	78.54	15.74
BMI	**25**	**4.6**
Riders	24	3.5
Non-riders	26.2	5.5

No statistically significant difference found in any of the anthropometric variables

Although horseback riders (10 as compared with 4 non riders) were somewhat more likely to be injured due to the nature of the sport they had been involved in (mostly broken bones in arms or legs), there was no significant difference found between the two groups ($p>0.05$).

POSTURAL ANGLES

The postural angles of the two groups differed with regards to cranio-cervical angle, neck flexion angle, acromio-vertebral angle and pelvic tilt (see Table 4). Riders' postural angles were evidently smaller indicating that their postural alignment in areas mentioned compared to the non-rider group.

Table 4: Postural angles in degrees for the rider and non-rider group as well as for normative data sourced from various studies

Angle	Group	Mean	Std.Dev.	t-test p value
H.F	Riders	75.9	9.14	
	Non-Riders	72.3	5.78	0.141
Literature	Straker et al.	71.0	11.1	
GA	Riders	15.25	7.52	
	Non-Riders	17.79	3.9	0.181
Literature	MacEvoy&Grimmer	13.1	9.0	
CCA	Riders	144.22	7.69	
	Non-Riders	151	10.55	0.020*
Literature	Straker et al.	160.8	12.7	
NF	Riders	39.66	5.61	
	Non-Riders	44.18	8.46	0.043*
Literature	Straker et al.	51.8	8.6	
CVA	Riders	49.83	5.69	
	Non-Riders	46.31	5.8	0.123
Literature	Gillian M.J.	49.58	4.1	
NA	Riders	127	6.21	
	Non-Riders	129.25	7.05	0.279
Literature	Burgess-Limerick	120.5	7.0	
AVA	Riders	28.74	9.81	
	Non-Riders	49.92	13.48	p<0.001**
Literature	Lewis et al.	57.4	9.6	
PT	Riders	14.23	6.01	
	Non-Riders	22.09	5.82	p<0.001**
Literature	Straker et al.	9.7	16.4	
ST	Riders	8.04	1.66	
	Non-Riders	8.24	1.08	0.789
Literature	Cureton K.J.	Direct comparison		

<u>Coding:</u> H.F (head flexion), G.A. (gaze angle), C.C.A. (cranio-cervical angle), N.F. (Neck Flexion), C.V.A (cranio-vertebral angle), N.A. (neck angle), A.V.A (acomio-vertebral angle), P.T. (pelvic tilt), S.T. (shoulder thrust)
Shadowed areas indicate previous studies' results in the area;
**. Variable is significant at the 0.01 level (2-tailed).
*. Variable is significant at the 0.05 level (2-tailed).

BERTOTI'S (OBSERVATIONAL) POSTURE ASSESSMENT SCALE

The postural assessment scores are presented in Table 6. The riders had a significantly higher mean score than the non-riders in terms of the shoulder, indicating that the position of the shoulder in the riding group was better (the score closer to 3).

Table 6: Bertoti's postural assessment scores

Area	Group	Mean	Std. Dev.	t-test P value
H.N.	Riders	1.43	0.90	0.055
	Non-riders	1.9	0.64	
S.H.	Riders	1.7	0.7	0.007*
	Non-riders	1.2	0.41	
S.P.	Riders	1.91	0.9	0.227
	Non-riders	1.6	0.75	
P.	Riders	1.65	0.83	0.066
	Non-riders	1.25	0.55	
Tot.	Riders	6.61	2.33	0.261
	Non-riders	5.9	1.62	

Coding: H.N. – head and neck, S.H. – shoulder, S.P. – spine, P. – pelvis
Gr. – group, R – riders, N – non-riders, Tot. – total score
*. Variable is significant at the 0.05 level (2-tailed).

DISCUSSION

The main hypothesis of this research study was partially supported. Importantly, the riders and non-riders differed on range of postural measures. Riders had better 'head and neck' and pelvic posture. As shown on the Bertoti's assessment, riders also had better shoulder posture. There was no statistical anthropometrical, age range and personality difference between riders and non-riders and the confounding factors of psychosocial and socio-economic profile were eliminated. Moreover, there were no substantial methodological differences between the present study and literature sources using the same methods except when concerning acromio-vertebral angle and shoulder thrust. For example, the author used the mastoid process as one of the markers instead of tragus of the ear that has been employed in previous studies.

In conclusion, horseback riding has the potential to positively influence the postural configuration of the human body. Further studies using similar methods but with larger samples are required to substantiate the value of horseback riding for correct posture.

REFERENCES

Ankrum, D. R. & Nemeth, K. J. (2000). Head and neck posture at computer workstations: What's neutral? Proceedings of the 14th Triennial Congress of the International Ergonomics Association, 5, 565-568.

Bertoti, D. B. (1988). Effects of therapeutic horseback riding on posture in children with cerebral palsy. Physical Therapy, 68(10), 1505-1512.

Brock, B. (1989). Therapy on horseback: Psychomotor and psychological change in physically disabled adults. Paper presented to The American Camping Association National Conference Seattle, Washington, 2-18.

Burgess-Limerick, R., Plooy, A., Fraser, K., Ankrum, D.R. (1999). The influence of computer monitor height on head and neck posture. International Journal of Industrial Eronomics, 23, 171-179.

Darling, D.W., Kraus, S., Glasheen-Wray, M.B. (1984). Relationship of head posture and the rest position of the mandible. The Journal of Prosthetic Dentistry, 52(4), 111-115.

Debuse, D., Chandler, C., Gibb, C. (2005). An exploration of German and British physiotherapists' views on the effects of hippotherapy and their measurement. Physiotherapy Theory and Practice, 21(4), 219-242.

Ferrari, R. & Russell, A. S. (2003). Regional musculoskeletal conditions: Neck pain. Best Practice and Research Clinical Rheumatology, 17(1), 57-70.

Gillian M.J. (1998). The correlation between surface measurement of head and neck posture and the anatomic position of the upper cervical vertebrae. SPINE, 23(8), 921-927.

Gough, B. J. (1993). Riding therapy: Is riding therapy a valid adjunct to more conventional physiotherapeutic techniques in the rehabilitation of the disabled patient or is it merely a leisure activity? Bradford and Airedale College of Health Physiotherapy Studies, 1-36.

Haehl, V., Giuliani, C., Lewis,C. (1999). Influence of hippotherapy on the kinematics and functional performance of two children with cerebral palsy. Pediatric Physical Therapy, 11, 89-101.

Harrington, G. (1999). Straight from the horse's mouth. The Therapist, 6(1), 29-31.

Harrison, A.L., Barry-Greb, T., Wojtowicz, G. (1996). Clinical measurement of head and shoulder posture variables. Journal of Orthopaedic and Sports Physical Therapy, 23(6), 353-361.

Heipertz, W. (1981). Therapeutic Riding: Medicine, Education, Sports. Ottawa: Greenbeit Riding Association for the Disabled, 16-31.

Hornacek, K. (2005). Basic, yet overlooked factors influencing posture in horseback-riding therapy: Hippotherapy with babies and toddlers, Scientific and Educational Journal of Therapeutic Riding, 1, 11.

Ionatamishvili, N., Tsveravaq, D., Avaliani, L. (2001). Riding therapy rehabilitation methods of child cerebral palsy. Scientific and Educational Journal of Therapeutic Riding, 7, 14.

Johnson, M.G. (1998). The correlation between surface measurement of head and neck posture and the anatomic position of the upper cervical vertebrae, SPINE, 23(8), 921-927.

Klassen, R. M. & Lynch, S. (2007). Self-efficacy from the perspective of adolescents with LD and their specialist teachers. Journal of Learning Disabilities, 40(6), 494-507.

Kendall, F.P., McCreary, E.K., Provance, P.G., Abeloff, D.K. (1993). Muscles testing and function. 4th ed. Baltimore: Williams & Wilkins.

Kyrklund, K. & Lemkow, J. (1998). Dressage with Kyra. Vermont, Trafalgar Square Publishing

Land, G., Errington-Povalac, E., Paul, S. (2001). The effects of therapeutic riding on sitting posture in individuals with disabilities. Occupational Therapy in Health Care, 14(1), 1-11.

Lewis, J. S., Wright, C., Green, A. (2005). Subacromial impingement syndrome: The role of posture and muscle imbalance. Journal of Shoulder and Elbow Surgery Board of Trustees, 14(4), 385-392.

McDonnell, J. M. (2006). Cervical/thoracic myalgia due to forward head posture: Non-pathological. Structural Yoga Therapy Research & Case Study Paper, 1-47.

MacGibbon, N.H., Andrade, C., Widener, G., Cintas, H.L. (1998). Effect of an equine-movement therapy program on gait, energy expenditure, and motor function in children with spastic cerebral palsy: A pilot study. Developmental Medicine and Child Neurology, 40, 754-762.

MacEvoy, M.P. & Grimmer, K. (2005). Reliability of upright posture measurements in primary school children. BMC Musculoskeletal Disorders, 6(35), 1-10.

MacKinnon, J.R., Noh, S., Laliberte, D.,Allan, D.E., Lariviere, J. (1995). A study of therapeutic effects of horseback riding for children with cerebral palsy. Physical & Occupational Therapy in Pediatrics, 15(1), 17-34.

MacKinnon, J.R. & Ferreira J.R. (2002). The meaning of therapeutic horseback riding for children with cerebral palsy. Scientific and Educational Journal of Therapeutic Riding, 8, 1-15

Mann, G. (2005). Examples of good practice in equine assisted therapy. Hampshire: George Mann Publications

Nettle, D. (2007). Personality: What makes you the way you are? New York: Oxford University Press

Pulig-Schatz, M. (1992). Back care basics: A doctor's gentle yoga program for back and neck pain relief. California: Rodmell Press

Sakakura, M. T., de Barros Santos, R., Cyrillo, F. N., Perdigao, A. P., Torriani, C. (2006). Electromyography comparative analysis of lumbar erector muscle with a cerebral palsy patient performing different postures on horseback. Scientific and Educational Journal of Therapeutic Riding, 5, 38-43.

Silkwood-Sherer, D. & Warmbier, H. (2007). Effects of hippotherapy on postural stability, in persons with multiple sclerosis: A pilot study. Journal of Neurologic Physical Therapy, 31, 77-84.

Straker, L. M., O'Sullivan, B. P., Smith, A. J., Perry, M. C. (2009). Relationships between prolonged neck/shoulder pain and sitting spinal posture in male and female adolescents. Manual Therapy, 14(3), 321-329.

Strauss , I. (1998). Hippotherapy – its unique position within physiotherapy. In Therapeutic Riding in Germany: Selected contributions. Warendorf: Selbstverlag.

Swift, S. (1998). Centered Riding. London: Kingwood Press

Uchiyama, H., Ohtani, N., & Ohta, M. (2011) Three-dimensional analysis of horse and human gaits in therapeutic riding. Applied Animal Behaviour Science, 135(4), 271-276.

Watson, D.H. (1993). Cervical headache: an investigation of natural head posture and upper cervical flexor muscle performance. Grieve's Modern Manual Therapy, 24, 349-359.

Wham, J. & Cert, B. (1990). Rhythmic facilitation: The treatment of neuromotor disorders using the rhythmical action of a walking horse. Paper written to accompany a talk and a video presentation given at the New Zealand Association of Occupational Therapy Jubilee Conference.

Would, J. (1996). Study on posture and the development of balance in disabled riders. Scientific and Educational Journal of Therapeutic Riding, 3-9.

Would, J. (2002). Improvements in walking and functional ability in children with cerebral palsy following hippotherapy: Research results. Paper presented at the International Symposium "From Rehabilitation to Social Companionship: A Journey with the Horse." Amiens, France: University of Picardie.

Yoo, W., Cho, S.H., Jeon, H., Cynn, H., Choi, H. (2008). Effects of the height of ball-backrest on head and shoulder posture and trunk muscle activity in VDT workers. Industrial Health, 46, 289-297.

ImageTool was developed in the Department of Dental Diagnostic Science at The University of Texas Health Science Center, San Antonio, Texas by Dr. C. Donald Wilcox, DSc, Dr. S. Brent Dove, DDS Dr. W. Doss McDavid, PhD and David B. Greer.

ARTICLE THREE

Quality of Movement and Motor Problems in Patients with ADHD: Implications for a Therapeutic Riding Approach

Birgitte Hokkerup, Physiotherapist

Correspondence to:
Birgitte Hokkerup
5627 Jondal
Norway
Phone: +47 98 40 50 48
Email: birgitte.hokkerup@kvamnet.no

QUALITY OF MOVEMENT AND MOTOR PROBLEMS IN PATIENTS WITH ADHD: IMPLICATIONS FOR A THERAPEUTIC RIDING APPROACH

ABSTRACT

The main focus of this article is motor problems in patients with ADHD in two core areas: (i) inhibition problems which affect fine and gross motor adjustment and regulation of power and flow of movement; and (ii) problems with proximal stabilization of the trunk that leads to compensative activity and high muscular tone in key movement muscles to maintain alignment (Stray, 2009). Identifying how these specific motor problems may affect riding leads to the question of finding an appropriate practical approach that is helpful for the patient. The case presented here is an example of how to use the concept of Balance and Movement, originated by Susanne von Dietze, in therapeutic riding practice for patients with ADHD. It was presented in April 2012 at the IVX International Congress of Therapeutic Riding in Athens and in August 2012 as a case study for the D-course "Balance and Movement: Further Education for Physiotherapists" in Gothenburg, Sweden.

KEYWORDS

Therapeutic Riding; Horse-Assisted Therapy; Motor Problems; ADHD; Pain.

BACKGROUND

I am a physiotherapist and work in primary healthcare in a small community in western Norway where I have also been working with therapeutic riding (TR) for 12 years. In Norway therapeutic riding (TR) is the term used for the kind of horse-assisted therapy offered exclusively by physiotherapists.

I have offered TR to several patients with ADHD. The object of this article is to share the theoretical and practical approach I have taken in order to meet their challenges. I have had to find my theoretical base in research that had nothing to do with TR and from there integrate the findings into my clinical work.

I often observed impaired coordination, muscle tension, and difficulties with releasing muscle tension in my patients with ADHD. I rely on the traditions of the Scandinavian movement system Basic Body Awareness Therapy (BBAT) and of Norwegian Psychomotor Physiotherapy. In this tradition, impaired quality of movement is understood in the light of using compensatory muscles to make up for poor proximal stability. While proximal stability is maintained by muscles that are not controlled by willpower, the compensatory gross movement muscles are controlled by will and easier to recruit, but not constructed for stabilizing work over time. Free breathing, postural stability and awareness in all exercises are essential to change these imbalances and to keep this compensatory muscle work from turning into a pattern (Skjaerven, Gard, & Kristoffersen, 2003)

I tried to find literature describing the motor problems I observed in patients with ADHD using keywords such as "proximal stability" and "ADHD", and found the work of physiotherapist Liv Larsen Stray. Stray shares the understanding of quality of movement from a Norwegian Psychomotor Physiotherapy approach (Stray, 2009). Stray has worked with patients, mainly children, with ADHD for 30 years. I attended her courses and have since adopted her theoretical and practical framework.. She completed her doctoral dissertation on children with motor problems and ADHD (Stray, 2009) and her latest publication is on adults diagnosed with ADHD (Stray et al., 2013).

In children with ADHD, motor problems are not part of the diagnostic criteria in neither the Diagnostic and Statistical Manual for Mental Disorders, Fourth Edition Text Revision - DSM IV-TR (American Psychological Association, 2000), nor the International Classification of Disorders - ICD, (World Health Organization, 1992). However, Stray found that observing the children in natural settings showed that they were struggling with tasks requiring fine muscular adjustments, regulation of gross movement, and stabilization of the trunk. She also noted that while movements often started in a correct manner, the children encountered problems when the movements were continued over time. Standardized assessment tools like the Movement Assessment Battery for Children (M-ABC) are not specific enough to reveal these problems (Stray, 2009).

Stray (2009) designed an assessment battery, the Motor Function Neurological Assessment, wanting to make characteristic motor issues visible to parents and to teachers. She found two core areas of motor problems in patients with ADHD:

i. Inhibition problems that affect fine and gross motor adjustment and regulation of power and flow of movement; and
ii. Problems with proximal stabilization of the trunk that lead to compensatory activity and high muscular tone in gross movement muscles (like lattimus dorsi, iliopsoas, sacrospinalis and calf muscles) to maintain alignment.

Once parents and teachers had seen the children struggle with the movements in the assessment it was much easier for them to understand how such problems might complicate daily life activities and interaction with others. Stray thought that the children were unrightfully blamed for poor behavior, for not being cooperative, easily distracted and fidgety. She saw much of this behavior as the result of their physical discomfort (Stray 2009).

In her research on motor problems and pain in adults diagnosed with ADHD, Stray found the same motor problems in adults as in children. While pain is not an issue among children with ADHD, she found significant correlations in the adult group between total score of motor problems and pain level and pain location (Stray et al., 2013).

In riding, qualities of movement such as a stable upper body, a moveable pelvis, independent arms and legs and a gentle hand are desired. The rider should be supple, balanced and in rhythm with the horse (von Dietze, 2003). However, without a stable core it is not possible to have an independent seat on a horse. For the rider with ADHD the problem with proximal stabilization is therefore a central challenge, and problems with inhibition are a further complication.

The stable core also is the basis for the independent arms and moveable legs. However, the rider with ADHD can be caught in his/her compensatory muscle imbalances with restricted movement in the pelvis, hips and shoulders. Trying to correct posture will only increase tension, as the rider will use the already overloaded movement muscles to meet the correction.

The practical approach that I use for my riders with ADHD I found in Susanne von Dietzes' work. Her understanding of finding a new body balance and increased range of motion through movement fits right into the Scandinavian traditions to improve quality of movement mentioned above.

Von Dietze (2003) allows the rider to move in order to experience the rhythm and eventually find the balance and suppleness we look for. Corrections like "don't get so tense" are replaced by "things to do, not 'don'ts'". Movements to increase awareness of middle position of joints replace corrections of posture. Basic Body Awareness Therapy shares this approach (Skjaerven et al., 2003). Emphasis is on process and on new body awareness through movement.

This case study is an example of how I use my theoretical base, gained from Stray (2009), and how I have implemented it using the practical approach of von Dietze (2003) to the benefit of the patient.

Client Background

My client is a young woman, "Maria", 27 years old. She was diagnosed with ADHD by a psychiatrist at the age of 21. This was a short time after she had given birth to a son. The stress of sleepless nights with a baby increased her lifelong feeling of not being in control of reactions and emotions; she even feared that she could harm the baby. Maria's mother had been diagnosed with ADHD some time prior to Maria giving birth, and Maria wondered if her problems could have the same explanation, which, in fact, they had. Maria has a history of muscular tension and pain, anxiety and depression. As a child she felt different and lonely, and she struggled with physical activities demanding coordination in school. She also had learning difficulties. Riding became her favorite activity, the horse was good company and the stable a place she enjoyed spending time. She felt that a theoretical education was too demanding in the light of her difficulties with learning and attention, so she chose a practical apprenticeship as a glazier, but dropped out because of pain in neck and shoulders. Maria is currently unemployed and receiving social security. She takes care of the house, her husband, their young son aged 6 and two dogs. At the time of the intervention, Maria's mother and a younger sister (who also has learning and concentration problems) lived in a caravan in their yard and relied on Maria for support for their problems. Maria has tried ADHD medication but stopped because she felt alienated from herself. She is working actively with stress handling strategies, guided by a mental health professional (a psychiatric nurse).

Maria has owned her Norwegian Fjord horse for 9 years, ever since the horse was a foal. The horse plays an important part in all aspects of her life, both as an activity and as consolation. Much of her identity and self-esteem is centered on life with her horse and she handles panic attacks by going to the horse. Maria has been an ordinary independent rider since childhood. Riding instructors had always praised her posture on the horse, but commented on poor use of her leg and weight aids. Stagnation in her riding because of pain in shoulders and hips and difficulties using her aids frustrated her.

PROBLEM

I first met Maria when I conducted a body awareness training session with local riders. She was sitting up straight beautifully on her horse, and it was only when I introduced some leg movements that I became aware of her limitations; she could not move her legs or pelvis more than a tiny bit without getting cramps and pain in her hip flexors. Releasing the muscles in her lower back was not an option for her. In addition, movement in her upper body often caused pain in the hips. Poor proximal stability and associated compensatory muscle activity seemed to cause restricted respiration, reduced mobility in hip, pelvis and shoulders and to promote muscle pain.

STRONG POINTS

I rate Maria's experience with horses, riding abilities and balance as well as her determination to work towards her goals as her strengths.

WEAK POINTS

The tension Maria experienced in her body was so intense that she would often have trouble resting her head on the pillow when she went to bed. She had been locked in an ADHD typical pattern of compensatory muscle work for years, so the strain on her muscles had been substantial for a long time.

GOALS AND REASONS FOR CHOOSING THERAPEUTIC RIDING

There was mutual agreement about the potential of TR to help Maria. Maria's own goals were to increase awareness of her body while on the horse. She was also keen to enhance her riding ability by better use of aids. She also wanted to reduce her discomfort while applying these aids. Including goals related to horseback riding was very important to the client and different from ordinary physiotherapy.

The therapeutic goals were:

- I. Gradually extending range of motion without triggering pain
- II. Increasing flexibility in all of her body through movement;
- III. Increasing postural stability

These goals were important for improving her riding and all aspects of her life.

The Intervention

As I was aware that Maria had ADHD, I made a connection between this and her physical symptoms from the very start of our contact, and it guided my understanding and approach. Maria, however, had not made the connection between her physical symptoms and ADHD before,

Frequency and Duration of Sessions

During our first meeting in the body awareness training session I gave Maria some tools to start her process, like moving her pelvis more in rising trot in all directions, and focusing more on her ability to move to start loosening up gradually than on a perfect posture.

After some months she had already profited from working with her new tools and was more flexible. We had weekly sessions of 30-45 minutes for 16 weeks. Eight of these sessions were on the horse, and eight were conducted without the horse. She was given "homework" to do in everyday life situations as well as on an air cushion and Balimo-chair for her to integrate this new body awareness in all aspects of her life. Maria worked with this approach daily, both when riding and in other situations such as driving a car or doing housework.

We used Maria's own Fjord mare and ordinary riding equipment in individual sessions.

Initial Focus in Treatment

Working with focused attention during riding (and during other daily life activities) increased her awareness of her rigid habitual patterns which is the basis for change. Simply having an explanation for her problems was an important step for her. She had feared something more serious was wrong with her because of her years of pain and tension, and no one had been able to explain it to her previously.

We mainly used walking to increase awareness of movement and mobilizing of pelvis and hips, upper body and shoulders. Walking was also chosen for tonus regulation, for exercises to increase postural stability, and for suppleness and rhythm. Exercise focusing on keeping the pelvis in the horse's rhythm and the seat bones connected with the horse while giving rhythmical tasks for her arms and legs (e.g., moving hands in a juggling rhythm or moving ankles or lower legs) is an example of this. It enhances postural stability and increases awareness of rhythm (von Dietze, 2003). As postural stability got better her seat got more flexible and the quality and flow of the movements of arms and legs improved. For progression, thesame exercises were used in trot. Rising trot was a good way to connect into the rhythm of the horse and to play with movement of the pelvis. Changing between forward seat and sitting down both in walking and in trot in a smooth flow increased awareness of the difference between stability and flexibility of the pelvis (von Dietze, 2003).

Changes During Treatment

When Maria acquired a more flexible seat a new problem became visible: slouching of the upper body. "But I was never like this", she burst out. Maria had kept her upper body erect by hollowing her lower back. Once her lower back began to release and her pelvis moved along with the horse's movement, there was no other point of stability ready to take over keeping the upper body as aligned as before and she slouched.

There was, therefore, the need for increased focus on stabilizing exercises for her upper body while riding. We used more exercises in the forward seat, like moving arms forward alternately without losing the position or moving the pelvis in all directions while still keeping the upper body aligned over the horse. We worked on arm movements to establish appropriate conditions for shoulder stability and independent hands. Patting the horse on the left side of its neck with her right hand, while staying centered with her pelvis, mobilized her thorax and shoulders. Holding a big imaginary ball in front of her demanded stability. We also worked on the ability to lift her sternum a little bit without hollowing her back or drawing back shoulder blades, in order for her to feel how she lengthened her upper body without tension and without losing the connection between the horse's movement and her seat bones. These exercises helped her build upper body stability (von Dietze, 2003).

Results

Maria lives with her ADHD typical tensions and they increase in periods of stress, but during the TR process she has obtained increased postural stability, increased flexibility and ways to handle her tensions better. We have tried to replace stabilizing strategies by compensatory muscles with conscious movement strategies, and she has made full use of these strategies. Whenever she feels tension or pain, she moves in a conscious way which releases her tensions and helps her keep up her postural stability. Through this she is able to handle situations that used to cause stress and pain and made her feel out of control.

As a rider she is now in the position to explore her weight aids, is able to use her leg aids and also able to demand and support more collection from her horse. I have followed her for over a year after the intervention and with only a little help occasionally her progress has continued. She has had long periods without any pain at all. She is now competing in dressage with good progress and can train without pain even when learning new skills. She has also become a TR volunteer at the riding center. She experiences this as a step towards some kind of employment, although she still feels she has a long way to go in that respect.

Discussion

Although the movements of the horse alone can help a lot of riders achieve suppleness and good posture, it is my experience that the movements alone are not enough to accomplish this for riders with ADHD and inhibition problems. Riders with ADHD who have developed a strong pattern of compensatory muscle activity need an instructor or therapist who understands their specific challenges. Otherwise increased tension in already overloaded compensating muscles is to be expected. Riding with the right guidance represents a valuable training method to work with both proximal stability and increased flexibility, but the therapist must be able to detect the difference between contact with compensatory muscle action and contact with proper proximal stability.

For adults with ADHD pain and tension is a challenge beyond what can be observed among children. Psychiatric co-morbidity is found in up to 80 % of adults (Stray, 2013). According to Stray et al., (2013), many adults like "Maria" with ADHD report that their pain problems have not been adequately explained and that the pain has caused anxiety. My experience with Maria makes me expect that TR represents a valuable therapy for many of these adults. It would enhance their body awareness through the contact with the horse and give them tools like Maria received in order to cope with their motor problems. Further research would be necessary to explore my presumptions.

Conclusion

Finding the research of a physiotherapist who does not work with TR at all has been crucial for me to fully understand the challenges my patients with ADHD have and has helped me choose an appropriate approach. This is why I find it important to integrate the research of colleagues in our horse-assisted work and document its value in a TR context.

In this case study I have explored the clinical implications of a theoretical base. I have looked for an approach that makes the rider recruit the postural muscles and not follow her habitual pattern of recruiting the muscles easiest to contact, which are the gross movement muscles that have established the compensatory pattern. I found this in von Dietze's (2003) work. She gives the rider tasks that cannot be done without an automatic activation of postural muscles. She connects every exercise into riding and the rhythm of the horse (von Dietze, 2003). This gives the body a way to readjust from an inappropriate pattern and it opens for a change and a new strategy to cope with motor problems, tension and pain.

REFERENCES

American Psychiatric Association (2000). Diagnostic and statistical manual of mental disorders (4th ed., text revision). Washington, DC: Author

Skjaerven, L., Gard, G.,,& Kristoffersen, K. (2003). Basic elements and dimensions to the phenomenon of quality of movement. Journal of Bodywork and Movement Therapies, 7(4) 251-260.

Stray, L. L. (2009). Motor problems in children with ADHD and clinical effects of methylphenidate as assessed with the MFNU. Doctoral Dissertation: University of Stavanger

Stray L. L., Kristensen, Ø., Lomeland, M., Skorstad, M., Stray, T., Tønnesen, F.E., (2013). Motor regulation problems and pain in adults diagnosed with ADHD. Behavioural and Brain Functions, 9, 18

Von Dietze, S.(2003). Balance in movement; The seat of the rider. London: Trafalgar Square Publishing.

World Health Organization. (1992). The ICD-10 classification of mental and behavioural disorders: Clinical descriptions and diagnostic guidelines. Geneva: Author

ARTICLE FOUR

Hippotherapy: The Mysterious Art of Healing or the Art of Rehabilitation?

Jerzy Krężel, Polish Hippotherapeutic Association

Correspondence to:
Jerzy Krężel
Polskie Towarzystwo Hipoterapeutyczne
ul. Nowoursynowska 100,
02 – 797 Warszawa, Poland
Email: jerzy.krezel@hipoterapia.org

HIPPOTHERAPY: THE MYSTERIOUS ART OF HEALING OR THE ART OF REHABILITATION?

ABSTRACT

Hippotherapy, according to the definition used by the Polish Hippotherapy Association, is a form of medical treatment that might improve a patient's condition in physical, mental and social domains. It is obvious that if hippotherapists want to develop and promote this treatment strategy, some effort should be made to have it accepted and approved by the medical community as a complementary treatment option. For countries like Poland, this effort could help us promote the emergence of the official profession of hippotherapist. Undoubtedly, a reasonable and professional development of this treatment strategy needs to be conducted in accordance with the philosophy of science. Unfortunately, some papers and proceedings on the topic of hippotherapy give the impression that not all therapists share that opinion. While some papers present objective results of studies as well as discussion based on scientific knowledge, others feed us with fairytale speculations decorated with modern physics terminology. This leads to an erroneous increase in the significance of pseudoscientific publications. The goal of this text is to emphasize the difference between a scientific and a pseudo-scientific approach. Several hippotherapy articles are presented as an example of science. As a counter-example, some pseudo-scientific revelations are also introduced. By means of analysis and comparison, the author attempts to show the benefits of an objective, evidence-based evaluation and promotion of hippotherapy, as well as the losses resulting from pseudo-scientific or even fictitious digressions. After clear demarcation of the two opposing approaches, the author discusses guidelines pertaining to planning and conducting evidence-based study. It is the results of these studies that may, after statistical analysis, scientifically show whether hippotherapy as a treatment approach is 1) harmless (that is, bringing no danger to the patient/client), and 2) improves the patient's functioning.

KEYWORDS

Hippotherapy; Science; Pseudoscience

Introduction

There is relatively large number of definitions of hippotherapy, and the definition varies depending on country and organization. For example, according to the Polish Hippotherapy Association (n.d.): "Hippotherapy is the therapeutic activity, oriented to improve physical, emotional, cognitive and/or social functioning of a human being, with a specially trained horse, which is an integral part of the therapeutic process". A qualified hippotherapist is responsible for conduction of the treatment, according to the directives of the physician referring the patient to hippotherapy, as well as in cooperation with other specialists who treat the patient. In Poland, six months of specialized training, which is separated into 186 hours of theory and 80 hours of practical classes is required. It is reasonable to say that individuals trained in hippotherapy believe in functional improvement of the patient following hippotherapy (e.g. decrease in pathological muscle tone, creation of correct walking pattern, increase in self-confidence, decrease in emotional disorders, improvement of perception, stimulation of attention, thinking, speech or even social relations), since they have had the opportunity to observe such improvement first hand. Since the treatment is of significance to the medical field, efforts are undertaken in some countries to have hippotherapy recognized as a therapeutic treatment complementary to other medical treatments. In order to approach that goal, evidence of its benefits and of its harmlessness needs to be provided.

Publishing information about the efficacy of hippotherapy should be done with careful consideration, especially when there is an attempt to reach an audience outside of the hippotherapy field. Regardless of the form of publication, it is strongly suggested to use scientific support, that is, to conduct an evaluation of hippotherapy and to share the results in accordance with the philosophy of science. While the meaning of the term varies within various fields of science, in the case of therapies that are recognized as medical treatment, sharing of information should follow the Evidence Based Medicine (EBM) approach and should be scientifically supported with statistically analyzed results. In the next two sections, the difference between science and pseudoscience is emphasized in relation to hippotherapy, and the benefits and disadvantages resulting from use of either approach are discussed.

Science and Knowledge

Many efforts were undertaken in the past to define what science is and what it is not, and effort is surely made to define it today. The most widely known example of such efforts is the works of philosophers and scientists of the Vienna Circle, which was active in the beginning of the last century. A person can choose a definition based on which philosopher or scientist she/he appreciates the most, however, it needs to be a definition or approach in which evidence and objectivity is preferred over speculation. The criteria for the demarcation of science from pseudo-science evolved in time from verification to falsification – the last one is presently considered valid.

The Carnap criterion of verification states that if a theory can be supported by experiment, then it is true, and if it cannot be verified in an experiment, it is incorrect. In light of today's view on science is has become apparent that this criterion is rather unfortunate, as verification by evidence may address only a finite number of experiments, and does not say anything about potential observations that could negate the theory. The advances of applied sciences give continuously enhanced tools to observe phenomena that could not be observed in the past. Thus, the chance of conducting experiments that can negate well known and so-called proven theories is increasingly likely. The criterion of falsifiability, developed by Karl Popper, negates the criterion of verification, stating that a theory cannot be empirically proven to be true. What can be proven is whether a theory is false. The experiment whose results are in accordance with a certain theory does not prove it to be true in general, but opposing results state that the theory is false (Popper, 1994).

SCIENCE AND PSEUDOSCIENCE

Science and pseudoscience can be explained with the following statements (Hansson, 2008):

- Science creates theories based on objective, measurable data.
- Scientific theories must be openly published (magic and alchemy is non-scientific knowledge).
- Scientific theories provide means for prediction (e.g. mathematical model and its numerical representation). The more precise prediction tools the theory provides, the more susceptible the theory is to falsification, and the easier it is to conduct an experiment which will negate it.
- Prediction on the basis of scientific theory must be in agreement with valid theories and knowledge.
- Scientific theories must be possible to be falsified. This means it must be possible to conduct experiments that are in contradiction to the theory.

- Pseudo-science is able to explain every phenomenon/fact.
- Pseudo-science is based on speculations.
- Pseudo-scientific observers and researchers create facts that verify their theories.
- Pseudo-scientific statements are published with no empirical validation.
- Pseudo-scientific statements cannot be submitted to testing procedures.
- Pseudo-scientific statements are usually in conflict with theories experimentally verified.
- Pseudo-scientific researchers usually do not present evidences of their theories.
- There is lack of objective falsifications in pseudo-science.

Scientific research is usually conducted according to the following guidelines (Hansson, 2008):

- Conduct preliminary observations
- Build hypothesis coming from analysis of results of observations
- Conduct reliable experiment (or gather historical data)
- Analyze results of experiment and accept or reject hypothesis on the basis of experimental results
- Repeat the whole procedure over again

While pseudoscience aspires to be called science it generally does not fulfill basic rules of science. It usually uses scientific language and terminology, but its assertions do not result from scientific procedures and usually do not contain detailed information. Moreover, pseudoscientific statements are quite often in contradiction with general science principles. Examples of pseudoscience can be found in alchemy, astrology, biorhythms, numerology, radionics, telepathy, and ufology.

Science and Statistical Analysis

The fundamental role of statistics is to provide a set of methods and guidelines for making sense of properly gathered or generated data, that is, to analyze and draw conclusions from data in a formalized way. It is the formalization that standardizes development and evaluation of science.

Attention should also be given to the difference between descriptive and inferential statistics. Descriptive statistics gives an opportunity to describe the specific sample that is being studied. This may increase the awareness of variability within the sample but cannot be applied to a population outside of the sample. Inferential statistics gives tools for generalizing from a set of samples to a population and provides means to assess its reliability.

In order to conduct an evaluation of hippotherapy with valid conclusions, the following steps should be taken:

1. Understand the problem (subject of evaluation)
2. Select and define variables that are relevant
3. Select/design the means to measure or determine the values of these variables
4. Conduct careful collection of the data, minimizing the influence/interference of undesired factors
5. Analyze the data in a formalized manner
6. Interpret the data and assess the reliability of the interpretation.

Science in Hippotherapy and Hippotherapy in Science

Scientific Approach

There are several evaluation techniques that can be used in order to gather and process empirical data about the outcomes of hippotherapy. The most commonly used are case study, descriptive study, comparative research and correlational research.

Case study. Case studies are a form of descriptive research in which the patient and the history of treatment is introduced. Even if many approaches and methods are introduced in the case study in order to perform in-depth analysis of the treatment strategy, this research method does not produce information in terms of a general evaluation of hippotherapy and its effects.

Descriptive study. Descriptive study is a research approach in which large samples are used. Its main features are subject to quantitative description. Descriptive data is used to summarize a data set. In the case of hippotherapy, descriptive study explains the influence of a specific sequence of treatment on a homogeneous group of patients.

Comparative research. Comparative research compares quantitative differences of different issues. A comparative study could examine the impact of two different therapeutic activities on two homogeneous groups of patients, such as the influence of hippotherapy and non-horse physical therapy on patients with cerebral palsy. These numbers might reveal differences in treatment efficacy, but conclusions cannot be drawn until the data has been statistically analyzed. Statistical analysis provides a means of telling whether the quantitative difference is statistically significant, that is, if the difference really results from the compared treatment outcomes, instead of simply being a statistical error.

Correlational research. Correlational research allows the study of concurrence of certain parameters with the effectiveness of applied treatment, the relationship between variables.

Variables can be age, sex, education, lifestyle, addictions, frequency of therapy, or such. At first sight, the difference seen in numbers may not be clearly visible. It is the statistical analysis that tells the difference. An approach to using this evaluation technique with regards to hippotherapy could be to investigate the frequency of therapy (the number of therapy sessions) versus its intensity (the duration of each session).

A comparative research approach can be particularly useful for a reasonable propagation of hippotherapy as medical treatment. There would be a greater impact if a study were conducted as a blind experiment, which, in the case of hippotherapy may be quite difficult but not impossible.

A scientific approach to research does not necessarily mean that a new theory needs to be created, or that existing theories need to be provided to support the evaluation.

- Scientific description: explanation/description of chosen phenomena leading to understanding and ability to predict certain behaviors. This is what individuals trained in hippotherapy need to pay attention to.
- Building scientific theory: creation of new theories to support observed phenomena. This is something that the field of hippotherapy does not necessarily need. Theories are built to explain observed phenomena, not to create them; therefore they require a lot of observation and analysis, consuming a huge amount of effort that may not be worth it.

With this in mind, a few papers related to hippotherapy or medical treatments in general will be presented.

Although the following paper by Witt et al. (2006) is not on the topic of hippotherapy, it is quite illustrative of the kind of scientific approach that could be followed in hippotherapy research. In this paper the authors investigate the effectiveness of acupuncture as treatment method used with patients with lower back pain. The study was performed with a randomized control trial of 298 patients divided into three subgroups: group treated with acupuncture, group treated with meta-acupuncture (more or less "fake" treatment – superficial needling at non-acupuncture points) and a group on the waiting list (control). The experiment was conducted as a blind trial – the patients did not know which of the first two groups they belonged to. The treatment, in both cases was done by experienced physicians who have specialized in acupuncture treatment. The outcome variable was the low back pain intensity given in the range of 0 to 100. The experiment was conducted during 8 weeks where each patient received 12 sessions. The results in numbers show that the actual acupuncture treatment was better than no treatment (group on waiting list), which is not surprising. A difference in numbers in favor of actual acupuncture (when compared to the fake treatment) was observed, although it was not as big as expected. The visible difference in numbers should not be the basis for drawing conclusions from the experiment. The authors processed the evaluation results statistically, revealing that this slight difference in favor of actual acupuncture was statistically insignificant. It is statistical significance that creates outcome in scientific evaluation. Not only is the paper a good example of a scientific approach, but it also presents the study in a way that is very clear, easy to read and easy to understand, because:

1. It clearly presents the background and the goal of the experiment.
2. It clearly describes the details of how the experiment was conducted.
3. It clearly shows methods used for conduction of the experiment as well as methods used to process the data.
4. It clearly shows the results, statistically processed, and conclusions.
5. Extensive data and details given in the paper are sufficient to repeat the experiment elsewhere, by other researchers.
6. It does not give a theory to support the evaluation, nor does it present a new theory. It presents the analyzed statistics in accordance to Evidence Based Medicine principles.

Although the paper was published with the goal of questioning the unclear efficacy of a treatment used for centuries, it may encourage us to do somewhat similar evaluation of hippotherapy, as compared to non-hippotherapeutic treatment. If the results show statistical significance in favor of hippotherapy, there will be evidence that can be presented to the medical field.

Another good example of a scientific approach is a hippotherapy-relevant paper by Muñoz-Lasa, Ferriero, Valero, Gomez-Muñiz, Rabini, & Varela, 2011. In the paper the effect of therapeutic horseback riding on balance in patients with multiple sclerosis was investigated. The authors compare the effectiveness of therapeutic horseback riding to traditional physical therapy. They also share all the details needed to perform such a study in other therapeutic centers and therefore their results can be verified, questioned and discussed. Although there are papers showing advantages of hippotherapy and therapeutic horseback riding over physical therapy, the authors found a niche, as there is only a limited number of publications dealing with the influence of therapeutic horseback riding on gait and balance of patients with multiple sclerosis. In this case, a blind trial was impossible, since the two therapeutic techniques are very different. The blind trial was in fact not needed, as the measurement tools (Extended Disability Status Scale (EDSS), Barthel Index, Tinetti Performance-Oriented Mobility Assessment (POMA), and especially gait analysis to assess spatio-temporal parameters and ground reaction forces) were far less subjective than in the case of the paper by Witt et al (2006). Results of this experiment clearly show statistically significant advantage of therapeutic horseback riding over physical therapy, in the case of balance improvement in patients with multiple sclerosis.

Very interesting and completely scientific papers have been published by a research group from Kitasato University (Japan), where the unquestionable value of a fully quantitative (dealing with properties measured and processed with numbers), rather than a qualitative (dealing with categorical properties which have no objectively measurable numeric characteristics) approach is shown (Matsuura et al., 2010; Matsuura, Ohta, Ueda, Nakatsuji, & Kondo, 2008; Matsuura, Masumura, Chiba, Nakatsuji, & Kondo, 2005; Matsuura et al., 2011).

The authors studied physical quantities and made conclusions on the basis of statistically processed results of their experiments. They investigated:

- The rhythmical movements of horse and human by quantitative measurements of acceleration of certain body parts
- The parameters of the horse's gait and therefore its influence on the patient, depending on its position in the therapeutic group (leader or follower)
- The relation between the conformation of the horse and its therapeutic properties
- A comparison of therapeutic horse riding to exercises on a horse simulator

All these research studies, conducted in accordance with philosophy of science, provide valuable and quantified knowledge of things that seem to be known to individuals practicing hippotherapy based on their experience with horses and their knowledge of therapy. Even if this non-scientific 'knowing' is correct, it is quantified findings that should be published in order to gain objective acceptance and appreciation.

In the case of medical and therapeutic treatment, there is a large number of well-developed and validated tests as well as methods of results analysis widely approved by the medical community that can be used. The efficacy of treatment must not be judged by eye; whenever it is possible quantitative evaluation and statistical analysis of data needs to be done. If it is not possible, qualitative tests, and specifically those that are officially considered valid in the medical community, should be used.

Finally, if this is not possible, testing procedures specifically for hippotherapy should be designed and based on scientific knowledge and the philosophy of science. As means of detecting and measuring various quantities improve (which includes new technology for sensors), new possibilities arise. A few things to consider concerning evaluation of certain physical and mechanical dysfunctions:

- What is the main variable allowing us to gather quantitative data for further processing and conclusions (see Matsuura et al., 2010; Matsuura, Ohta, Ueda, Nakatsuji, & Kondo, 2008; Matsuura, Masumura, Chiba, Nakatsuji, & Kondo, 2005; Matsuura et al., 2011)?
- What quantity is to be measured (for example, angular displacement of thigh or muscle tension)?

- What tools will be used to measure selected quantities and what are their measurement properties (sensitivity, measurement error margin)?
- How often should measurements be taken and what is the number of measurements in a series (statistical analysis usually requires a large number of measurements)?

Experiments need to be very well described in order to be easily replicable. If the experiment procedures, conditions or any other parameters are hidden, then, according to the philosophy of science, it is no longer science. In that case, even if detailed physical or medical explanations with well-defined mathematical models are presented, they are useless because the whole undertaking cannot be repeated and therefore it cannot be verified. In the case of novel issues and theory, it also cannot be falsified. The publications presented in this section can help bring hippotherapy closer to being recognized as a valid form of rehabilitation.

Pseudoscientific Approach

Although against the art of scientific writing, I have decided that references concerning pseudoscientific revelations will be omitted in order to neither insult anyone nor to bring about ridicule.

There as some papers and even books where authors deliberately omit scientific evidence of their revelations and explain that the reason for this is that the common hippotherapy practitioner has such poor knowledge of research and related terms that it would not be necessary. Examples of these revelations are:

- Holography as a way to describe how horses influence humans
- The impact of morphogenetic fields on human-horse interactions
- Chakras and energetic points
- Auras as a visible illumination of the mood and condition of the patients and the horse. This aura shows how the horse absorbs "negative energy" from the patient and heals him/her. Moreover, thanks to aura observation, it can be seen how this negative energy is injected to the paddock ground, when the horse "shakes out" after a therapy session.
- Kirlian photography as a photographic means of capturing auras (even though some sources say that an aura cannot be captured)
- Superluminar connections, infrared radiation, and similar methods used to describe the way the horse affects the patient

Authors of these supposed scientific findings do not seem to mind that:

- Holography is a means to record and reconstruct the intensity and phase of a light field, producing an image of an object after the object is no longer present.
- Morphogenetic fields involve a belief that there is a field of an unspecified nature that affects live beings.
- Aura is a kind of discharge that can allegedly be seen by gifted people. It cannot be treated in a scientific manner because it is not something that can be proven, but it can only be believed in.
- Kirlian photography is a means of registering the corona discharge, a well known and widely described physical phenomenon,
- Superluminar speed and infrared radiation belong to the domain of physics and optics in particular.

Including pseudoscientific concepts in hippotherapy-related writing could negatively affect the view of hippotherapy in the medical field, and the authors of these texts do not seem to be aware of this.

CONCLUSION

The future of hippotherapy relies on it being considered among objectively investigated medical and therapeutic treatments, not among the miraculous and mysterious healing practices. If most hippotherapy practitioners and researchers would agree to accept and follow a scientific approach in its evaluation, more sophisticated research challenges would arise and could be addressed.

In many fields of science, especially where qualitative evaluation or quantitative measurement is required, much effort is put towards developing common guidelines or directives concerning the conduct of an experiment. This process, called standardization, is also needed in hippotherapy, in accordance with the rules of science. The more issues we have standardized, the more common language we speak and thus, the more informative the dissipation of our knowledge. It is up to practitioners and stakeholders to decide whether hippotherapy is a mystical art of healing or a solid art of rehabilitation. It is also up to the field of hippotherapy to decide how non-medical and medical professions view this treatment strategy.

HIPPOTHERAPY: THE MYSTERIOUS ART OF HEALING OR THE ART OF REHABILITATION?

I, as a hippotherapy practitioner, would like to encourage you to believe in experiments, numbers and mathematics. Why? Because it is a scientifically reliable way of propagating hippotherapy. If our studies are properly designed and conducted in accordance with a scientific approach, the evidence presented will be very hard to deny.

Instead of trying to attach spectacular physical theories to hippotherapy, we could carefully perform a series of laborious quantitative measurements (or at least qualitative evaluation) and process the data statistically in order to be able to say that this treatment strategy is more effective than traditional physical therapy in cases of muscle contracture, for instance, with a probability of x and with statistical significance. This will be received differently by the medical community than something like "because of the impact of energy fields, the horse takes bad energy from the patient and heals him/her".

References

Hansson, S. (2008). *Science and Pseudo-Science*. Stanford Encyclopedia of Philosophy.

Matsuura, A., Masumura, K., Chiba, Y., Nakatsuji, H., & Kondo, S. (2005). Analysis of rhythmical movements of both horse and the rider using accelerometer. *Journal of Japanese Society for Applied Animal Behaviour, 41*(1), 5-11.

Matsuura, A., Nagai, N., Funatsu, A., Irimajiri, M., Yamazaki, A., & Hodate, A. (2011). Trekking and exercising with a riding simulator on autonomic nervous activity. *Journal of the Interactions of People & Animals, 24*(1), 65-77.

Matsuura, A., Ohta, E., Ueda, K., Nakatsuji, H., & Kondo, A. (2008). Influence of equine conformation on rider oscillation and evaluation of horses for therapeutic riding. *Journal of Equine Science, 19*(1), 9-18.

Matsuura, A., Tanaka, M., Irimajiri, M., Yamazaki, A., Nakanowatari, T., & Hodate, K. (2010).Heart rate variability after horse trekking in leading and following horses. *Anim Sci J. 81*(5), 618-621.

Muñoz-Lasa, S., Ferriero, G., Valero, R., Gomez-Muñiz, F., Rabini, A., & Varela, E. (2011). Effect of therapeutic horseback riding on balance and gait of people with multiple sclerosis. *G Ital Med Lav Ergon, 33*(4), 462-467.

Polish Hippotherapeutic Association. (n.d.). *Canons of Polish hippotherapy.* Retrieved from: http://www.pthip.org.pl/hipoterapia/kanony_hipoterapii/

Popper, K. (1994). *Zwei Bedeutungen von Falsifizierbarkeit*. München: Deutscher Taschenbuch Verlag

Witt, C., Jena, S., Linde, K., Streng, A., Wagenpfeil, A., Irnich, A., Walther, H. U., Melchart, D., & Willich, S. (2006). Acupuncture in Patients With Chronic Low Back Pain. *Arch Intern Med, 166,* 450-457.

Made in the USA
Monee, IL
05 November 2026

67977841R00046